Microeconometrics and MATL

Microeconometrics and MATLAB®: An Introduction

Abi Adams

Damian Clarke

Simon Quinn

OXFORD
UNIVERSITY PRESS

OXFORD
UNIVERSITY PRESS

Great Clarendon Street, Oxford, OX2 6DP,
United Kingdom

Oxford University Press is a department of the University of Oxford.
It furthers the University's objective of excellence in research, scholarship,
and education by publishing worldwide. Oxford is a registered trade mark of
Oxford University Press in the UK and in certain other countries

MATLAB® and Simulink® are registered trademarks of The MathWorks, Inc.

For MATLAB and Simulink product information, please contact:
The MathWorks, Inc.
3 Apple Hill Drive
Natick, MA, 01760-2098 USA
Tel: 508-647-7000
Fax: 508-647-7001
E-mail: info@mathworks.com
Web: mathworks.com
How to buy: www.mathworks.com/store

First Edition published in 2015
Impression: 1

Published in the United States of America by Oxford University Press
198 Madison Avenue, New York, NY 10016, United States of America

British Library Cataloguing in Publication Data
Data available

Library of Congress Control Number: 2015939865

ISBN 978-0-19-875449-7 (hbk.)
        978-0-19-875450-3 (pbk.)

Printed in Great Britain by
Clays Ltd, St Ives plc

■ CONTENTS

▨ LIST OF FIGURES

■ LIST OF TABLES

■ PROLOGUE

SOCRATES: And now, I said, let me show in a figure how far our nature is enlightened or unenlightened:—Behold! human beings living in an underground cave, which has a mouth open towards the light and reaching all along the cave; here they have been from their childhood, and have their legs and necks chained so that they cannot move, and can only see before them, being prevented by the chains from turning round their heads. Above and behind them a fire is blazing at a distance, and between the fire and the prisoners there is a raised way; and you will see, if you look, a low wall built along the way, like the screen which marionette players have in front of them, over which they show the puppets.

GLAUCON: I see.

SOCRATES: And do you see, I said, men passing along the wall carrying all sorts of vessels, and statues and figures of animals made of wood and stone and various materials, which appear over the wall? Some of them are talking, others silent.

GLAUCON: You have shown me a strange image, and they are strange prisoners.

SOCRATES: Like ourselves, I replied; and they see only their own shadows, or the shadows of one another, which the fire throws on the opposite wall of the cave?

GLAUCON: True, he said; how could they see anything but the shadows if they were never allowed to move their heads?

SOCRATES: And of the objects which are being carried in like manner they would only see the shadows?

GLAUCON: Yes, he said.

SOCRATES: And if they were able to converse with one another, would they not suppose that they were naming what was actually before them?

GLAUCON: Very true.

SOCRATES: And suppose further that the prison had an echo which came from the other side, would they not be sure to fancy when one of the passers-by spoke that the voice which they heard came from the passing shadow?

GLAUCON: No question, he replied.

SOCRATES: To them, I said, the truth would be literally nothing but the shadows of the images.

GLAUCON: That is certain.

SOCRATES: And now look again, and see what will naturally follow if the prisoners are released and disabused of their error. At first, when any of them is liberated and compelled suddenly to stand up and turn his neck round and walk and look towards the light, he will suffer sharp pains; the glare will distress him, and he will be unable to see the realities of which in his former state he had seen the shadows; and then conceive someone saying to him, that what he saw before was an illusion, but that now, when he is approaching nearer to being and his eye is turned towards more real existence, he has a clearer vision,—what will be his reply? And you may further imagine that his instructor is pointing to the objects as they pass and requiring him to name them,—will he not be perplexed? Will he not fancy that the shadows which he formerly saw are truer than the objects which are now shown to him?

Plato*

---

* Plato in Rouse W. H. D. (editor), *The Republic Book VII*, Penguin Group Inc. (1961), pp. 365–401.

■ INTRODUCTION

As economists, we want to learn about human behaviour—that is why we build models, collect data, and care about econometrics. This is a book about some of the techniques that economists use to create a direct dialogue between economic theory and econometric estimation. These techniques are designed to help us think about the incentive structures that agents face and the implications of those structures for empirical observation—the 'shadows cast' by incentives upon the variables that we observe.

This book is designed as a practical guide for theory-based empirical analysis in economics. We are concerned with data that has been collected from individuals: individual households, individual firms, individual workers, and so on. This is what is sometimes termed 'microdata'. Specifically, then, this is a book about fitting microeconomic models to microdata, to improve our understanding of human behaviour.

MATLAB® is *not* the central plot of this story—though it is certainly the lead character. Our goal in this book is not to provide a comprehensive introduction to MATLAB. There are already plenty of good books available to do that.[1] Further, our goal in this book is not to teach you microeconometric theory. Rather, our goal is to discuss a series of standard problems in applied microeconometrics, and show how you can use MATLAB to tackle each of them.

It is quite unlikely that *any* of the specific models that we study here will fit perfectly any particular empirical problem that you face in your own work—but that, in a sense, is exactly the point. There are many excellent textbooks that cover standard microeconometric methods, and several excellent software packages for implementing those methods—often requiring just a single line of code for any given estimator. Of course, all of these methods can be implemented in MATLAB, but this is not where MATLAB's comparative advantage lies.

The beauty of MATLAB is its extraordinary flexibility. MATLAB allows us easily to build and adapt our own estimators. It thereby opens entire classes of new models—and, therefore, new ideas—that standard econometrics packages do not allow. Of course, when it comes to econometric algorithms, there will always be an important role for pre-bottled varieties off the shelf. But in this book, we will brew our own.

---

[1] For example, you could see Hahn and Valentine (2013).

Structure of the Book

We start in Part I, with topics that form the foundation of microeconometrics. After a brief review of basic MATLAB syntax in Chapter 1, we show how MATLAB can be used to solve some of the basic optimization problems that we encounter as economists. In Chapter 2, we use MATLAB to model the behaviour of optimizing agents. Here we will encounter the key functions `linprog` and `fmincon` for the first time—functions that will make repeated appearances throughout this book. In Chapter 3, we use MATLAB's optimization techniques for a different purpose: to find model parameters that best fit the data. This chapter introduces estimation by Maximum Likelihood and by Generalized Method of Moments.

With these important bases covered, we move to a series of applied topics. In Part II, we discuss discrete choice. Chapter 4 is about discrete multinomial choice. Here we introduce the concept of Maximum Simulated Likelihood— and, with it, the general notion of estimating by simulation. In Chapter 5, we turn to discrete games—that is, discrete choice problems among multiple players, in which each player's payoff depend upon the others' actions. This is an important form of discrete choice problem in its own right, but also provides a useful foundation for thinking generally about the numerical modelling of strategic interactions.

Part III is about time. In Chapter 6, we introduce decisions on a finite horizon. This framework is useful for understanding many important economic choices—such as investment in human capital. However, some economic choices—such as a firm's optimal investment decision—do not have a known end point. In Chapter 7, we discuss how to solve these models on an *in*finite horizon. We close the section by addressing the estimation of dynamic models.

Often, we have little understanding of the relationship between variables of interest. At other times, this relationship may be too complicated to model easily with standard parametric models. Part IV introduces nonparametric and semiparametric regression. In Chapter 8, we introduce kernel regression and other local regression estimators. In Chapter 9, we wriggle around the 'Curse of Dimensionality' by combining parametric and nonparametric methods.

Part V is about optimizing MATLAB code to run efficiently. For complex models, small tweaks in your code can result in large gains in efficiency. We will introduce MATLAB's Parallel Computing Toolbox, which allows you to easily run jobs across multiple processors.

Why MATLAB?

There is no shortage of advice about which language you should use for technical computing. Ask around. You will likely end up with a strange list of letters, names, and even an animal: C++, Fortran, GAUSS, GNU Octave, OxMetrics, Java, Julia, Perl, Python, R, S, Scilab, Stata . . . Often these well-meaning suggestions will be accompanied by deeply impassioned arguments about why this language is absolutely *the best* language to use for microeconometrics—and that anyone not using this language is a luddite. We will try not to be so dogmatic!

This is a book about MATLAB—but this does not mean that you should work only in this language. For example, Stata is an excellent package for a wide range of standard econometric methods. However, for writing and running *specialized* models, MATLAB is a great choice. With its extraordinary flexibility, MATLAB allows us easily to build and to adapt our own estimators. In this way it opens entire classes of new models—and, therefore, new ideas—that many econometrics packages do not allow.

In particular, here are some of the reasons why we love using MATLAB for our own work:

1. Being based around matrices, MATLAB code takes a similar structure to the textbook formulae with which we are familiar. Converting estimators from the page to the computer is, therefore, a comfortable step.

2. Many auxiliary routines that could take weeks to set up in other languages are readily available in MATLAB. These include packages for parallel computing, for numerical optimization, and platforms allowing integration with other languages. In this textbook, we will encourage you to code your own estimators but there is no reason to code the auxiliary routines yourself!

3. It is very simple to get MATLAB up and running. There is no need to install compilers, special text editors, different operating systems, and so forth. Indeed, MATLAB will already be installed on computers where you work, and you can get started just by pointing and clicking.

4. There is a large community of MATLAB programmers—which means that many great resources are freely available online, and that many of your friends and colleagues will use this language. This is a really important consideration when choosing a language! It is so practical—and fun—to be able to resolve problems with friends.

So let the journey begin. We hope that the concepts and techniques covered in this book will open new possibilities for you as an applied researcher.

Abi, Damian, and Simon

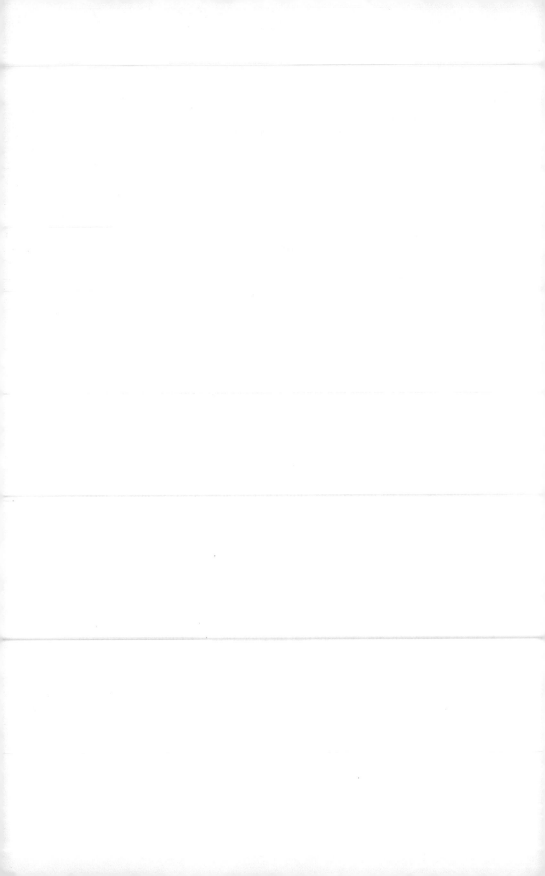

Part I
Foundations

1 Entering the 'Matrix Laboratory'

> The limits of my language mean the limits of my world.
>
> Wittgenstein*

MATLAB is a computer language for doing maths. Its name is short for 'matrix laboratory', and its purpose is simple: to provide a very powerful and very flexible way of solving mathematical problems. In this chapter, we will run a series of exercises to illustrate the simplicity with which MATLAB handles matrices—and to discuss good coding techniques for doing so. This will provide a foundation for the more complicated concepts and structures that we will cover later. We recommend that you read this book with MATLAB open in front of you and run the commands yourself as you encounter them in the text. We have found learning-by-doing to be the best way (and the most enjoyable!) of getting to grips with a new programming language and new econometric techniques.

In this book, we assume that you have a working knowledge of the MATLAB interface and can navigate your way between the different components of the MATLAB desktop. If this is your first time using the program, there are a number of excellent books and online resources to bring you up to speed. For example, Hahn and Valentine (2013).

The simplest way to interact with MATLAB is through the 'command line', and this is where we will begin. The command line operates like a calculator. We can see this by a none-too-complicated calculation:

```
>> 1 + 1
ans =
     2
```

We can use the command line to create matrix variables to store our results. Let's start with a simple variable, y:

```
>> y = 1 + 1
y =
     2
```

---

* *Tractatus Logico-Philosophicus*, Ludwig Wittgenstein, Copyright (1975) Routledge, Reproduced by Permission of Taylor & Francis Books UK.

As its name suggests, MATLAB is designed to deal with matrices very simply and effectively; in MATLAB, we can enter any variable as a matrix, simply by using commas to separate columns and semi-colons to separate rows. For example, let's create a simple 3×2 matrix (which we will call 'x'), and then multiply that matrix by two:

```
>> x = [1, 2; 3, 4; 5, 6]
x =
     1    2
     3    4
     5    6

>> 2*x
ans =
     2    4
     6    8
    10   12
```

We can check which matrices are stored in memory by using the commands who (for a short summary) and whos (for a longer summary):

```
>> who
Your variables are:
ans   x    y
>> whos
  Name       Size        Bytes  Class      Attributes
  ans        3x2            48  double
  x          3x2            48  double
  y          1x1             8  double
```

MATLAB reports that we have three matrices in memory: ans, x, and y. You should not be surprised to see x and y in memory; we just created these matrices, and we can check their contents simply by entering the matrix names at the command line:

```
>> x
x =
     1    2
     3    4
     5    6

>> y
y =
     2
```

The matrix ans may be more confusing. This matrix stores MATLAB's most recent answer that has not been stored in any other matrix. If we enter ans at the command line, we will ask MATLAB to recall its response to our earlier expression '2*x':

```
>> ans
ans =
        2       4
        6       8
       10      12
```

Notice that if we enter another expression that is not assigned to any other matrix, MATAB will use ans to store this new expression:

```
>> 5 * 5
ans =
     25

>> ans
ans =
     25
```

MATLAB has a very large range of mathematical operators. Our goal here is *not* to provide a comprehensive discussion of these. MATLAB provides excellent help files and a large range of online resources, and we do not want to use this book to describe what is available elsewhere. For example, to learn about MATLAB's arithmetic operators, you can simply search online to find the relevant help page.[1] To learn the syntax of a particular command, we can use MATLAB's extensive help documentation from the command line:

```
>> help ones
```

Instead of discussing an ungainly list of commands and operations at this point, we will instead explore different techniques as they become relevant for our analysis of various microeconometric models. And so we begin—with an illustration of the most popular microeconometric technique of them all . . .

1.1 OLS in MATLAB: 'Hello, world!'

A simple way to become familiar with the basic workings of an econometric program is to run an Ordinary Least Squares regression. In some ways, this is the 'Hello, world!' of the applied researcher. 'Hello, world!' is the test program

---

[1] In this case: http://www.mathworks.co.uk/help/matlab/ref/arithmeticoperators.html.

which many computer programmers run when they first learn a language—
to discover its basic syntax, and to ensure that it is running correctly. Such
programs simply print the words 'Hello, world!' and then terminate.[2] While
the OLS regression requires a little more work than just printing a simple statement, it provides us with a good opportunity to work with the basic building
blocks of MATLAB.

Almost every applied research is familiar with Stata, and almost everyone
who is familiar with Stata has, at some point or another, come across the
auto.dta dataset. This is a dataset included by default when Stata is installed,
and contains data on a series of models of cars in 1978. We ask you to open Stata
briefly and, using the auto dataset, run a regression of mileage per gallon upon
the car's weight and price.[3] We will denote mileage per gallon by the $N \times 1$
vector y, and will use the $N \times 3$ vector X to stack values of (i) price, (ii) weight,
and (iii) the number 1.

Our OLS model is, of course:

$$y = X\boldsymbol{\beta} + \boldsymbol{\varepsilon}, \tag{1.1}$$

where $\boldsymbol{\beta}$ is a 3×1 vector of parameters. We denote the OLS estimate of $\boldsymbol{\beta}$ as
$\hat{\boldsymbol{\beta}}$; we can find $\hat{\boldsymbol{\beta}}$ straightforwardly in Stata . . .

```
. sysuse auto
(1978 Automobile Data)

. reg mpg price weight, noheader
```

mpg	Coef.	Std. Err.	t	P>|t|	[95% Conf. Interval]
price	-.0000935	.0001627	-0.57	0.567	-.000418	.0002309
weight	-.0058175	.0006175	-9.42	0.000	-.0070489	-.0045862
_cons	39.43966	1.621563	24.32	0.000	36.20635	42.67296

Let's now write these three variables to the file auto.csv:

```
. outsheet mpg price weight using auto.csv, nonames comma
```

To run the same regression in MATLAB, we first need to import the data in
auto.csv. Before being able to import this data, we must ensure that our
current working directory contains the auto file. In order to move to this file,
we use the commands pwd (print working directory), cd (change directory)
and ls (list the contents of the current directory). After choosing the correct
working directory, we can import using dlmread:[4]

---

[2] In MATLAB, such a program would be quite simple, containing just disp('Hello, world!').

[3] And apologetically, for any readers expecting that this book would be based entirely on MATLAB . . .

[4] MATLAB requires a straight apostrophe (which looks like this: ') around the file name. If you enter the typical single quote operator (which look like this: ') you will find that MATLAB will not allow you

```
>> DataIn    = dlmread('auto.csv');
>> y         = DataIn(:,1);
>> X         = DataIn(:, 2:3);
>> size(X)

ans =

    74    2

>> X         = [X, ones(74,1)];
```

The above block of code involves various new commands. Check that you can run each command without problems in your MATLAB window now. Try running each command without the semi-colon at the end of the line; this will allow you to see the full output each time. The most important command is dlmread, which reads in the data from auto.csv. The entire dataset is stored as a matrix named DataIn. If you want to check that auto.dta has imported correctly, you can use the command whos('DataIn') to see the details.

You can now see how MATLAB is structured around matrices. In the first line of code, we store the data as a 74 × 3 matrix, which we then manipulate into a vector of the dependent variable y (mpg) and a matrix of explanatory variables X (adding a vector of ones for the constant). It is worth noting here that the notation DataIn(:, 1) implies that we take data from every single row (':') of column 1 in matrix DataIn.

The previous section of this chapter showed that we can enter matrices by hand at the command line (parsing with commas and semi-colons), but we will rarely need to do this. Generally we will either read in data directly as a matrix (as we have done here), or will use MATLAB's matrix-based operations to simulate data from economic models.

Now that we have two matrices (X and y) that contain the relevant data from Stata's auto dataset, we can run our regression. This requires little more than introductory econometrics, namely the formula:

$$\hat{\beta} = (X'X)^{-1}(X'y). \tag{1.2}$$

MATLAB's syntax follows Equation 1.2 closely. The only specialized function that we require is inv, which allows us to invert the $X'X$ matrix:

```
>> XX=X'*X;
>> Xy=X'*y;
>> BetaHat=inv(XX)*Xy
```

to move on to the next line at the command prompt. If this happens, you can break out of a half-written command using Control-C.

```
BetaHat =

    -0.0001
    -0.0058
    39.4397
```

Here we have calculated our coefficient matrix `BetaHat`. You will notice that our result is equal to those coefficients which we calculated earlier in Stata.[5]

We may also be interested in ensuring that `BetaHat` is correct to more than four decimal places. We can do this by changing MATLAB's output format to the *long* format, (» `format long`) which displays up to fifteen digits for 'double' variables.[6] Try changing the format and then displaying `BetaHat`. To change back to the traditional output format, just enter `format` once again.

Now that you have experienced MATLAB's functionality using OLS, you can practise what you have learned with an extension in Exercise (i) at the end of this chapter. Here, we ask you to calculate the standard errors of the coefficients estimated in the previous regression.

1.2 The Beauty of Functions

So far, our analysis has simply involved typing instructions into MATLAB's command line. This is effective but not very efficient. What we need is a method of saving commands so that we can run them later—and, if necessary, run them many times, with different data, different parameters, and different options. In MATLAB, we can do this with an 'M-file'. We create M-files through MATLAB's Editor window, which we can access by typing `edit` at the command line. M-files are to MATLAB what do-files are to Stata.

The most useful application of an M-file is to define a *function*.[7] In MATLAB, a function is a special type of program. There are three main elements to a function:

(i) Functions accept *inputs*.
(ii) Functions return *outputs*.

---

[5] Note that there is nothing (except perhaps a desire for clear exposition) that stops us from calculating $\hat{\beta}$ in a single step. This would look like: `BetaHat=inv(X'*X)*X'*y`, or alternatively using the functionality of MATLAB's backslash (`mldivide`): `BetaHat = (X'*X)\(X'*y)`.

[6] We resist the temptation to dive into a tangential discussion about the different levels of precision with which MATLAB can store numbers. You can look this up by searching for concepts like 'double' and 'single'.

[7] We can also use an M-file to define a *script*, but we will not spend much time discussing scripts.

(iii) Each function is *self-contained*; this means that each function can access *only* those variables that are passed to it as an input, and can store variables *only* through returning them as outputs.[8]

To get to grips with functions, let's return to the OLS regression that we ran earlier.[9] A regression is a perfect candidate for a function; each OLS regression is computationally equivalent but the inputs and outputs for each regression will vary depending upon the X and y variables that you are analysing. Therefore, you might be interested in permanently having a function available that you can call to calculate regression results. This is useful to save time when typing in commands at the command line, and to limit careless mistakes from typing in the calculation of β many times.

Here is an example of a function that we have written, called OLS. You should be able to open this file in the MATLAB editor by opening the file OLS.m. You can either open this file by using MATLAB's drop-down menus, or alternatively, by typing edit OLS.m at the command line.

OLS.m

```
 1  function [Beta, se] = OLS(y,X)
 2  %-----------------------------------------------------
 3  % PURPOSE: performs an OLS regression
 4  %-----------------------------------------------------
 5  % INPUTS: y: N-by-1 dependent variable
 6  %         X: N-by-K independent variable
 7  %-----------------------------------------------------
 8  % OUTPUT: Beta: OLS coefficient vector
 9  %         se: standard error of beta
10  %-----------------------------------------------------
11
12  %----- (1) Calculate the coefficients -----------
13
14  Beta   =   (X'*X)\(X'*y);
15
16  %----- (2) Calculate the standard errors --------
17  yhat   =   X*Beta;
18  u      =   yhat - y;
19  N      =   length(y);
```

[8] Strictly speaking, a function could save a variable to a file on the disk—but this is an unusual exception to the rule, and not one that we will often want to use.

[9] OLS is a useful illustration of MATLAB's basic concepts and basic functionality. But we will leave these sorts of standard econometric applications after this chapter. In many respects, it is much easier to implement standard estimators in Stata—and, if you *would* prefer to use MATLAB, there is an extensive set of functions already available through the Econometrics Toolbox.

```
20  K       =   size(X, 2);
21  sigma   =   sum(u.*u)/(N-K);
22  v_mat   =   sigma * inv(X'*X);
23  se      =   diag(sqrt(v_mat));
24
25  return
```

There are a number of things worth highlighting here, either because they are required for the code to run, or because they are good practice when writing functions.

(i) The first line of the function tells MATLAB (a) the *name* of the function (OLS), (b) the *inputs* to the function (y and X), and (c) the *outputs* from the function (Beta and se). The first line shows the correct syntax for this; we always start a function by some version of:

$$\text{function } [output] = \text{function\_name}(inputs) \quad (1.3)$$

Critically, this does *not* mean that when we call the function OLS we must use variables named y and X. Instead, it just means that, *within the program*, the variables we have introduced will be *locally* referred to as y and X. This will become apparent when we run the function shortly.

(ii) There are various lines of text that immediately follow the first line, each of which is prefaced by the % symbol. MATLAB reads the % symbol as saying 'skip this line', which allows us to write comments in our code without interrupting the running of our script. In this way, the lines of comments can then be thought of as an explanation (either for other users, or for ourselves in the future—and we recommend being kind to your future selves!) to help understanding of our code. As an added benefit, the lines of comments that directly follow the function are used as the help file to the function. So, when we type help OLS at the command line, the output will remind us what we need to input, and what we should expect as output. Performing this for our OLS function returns:

```
>> help OLS
---------------------------------------------------------------
    PURPOSE: performs an OLS regression
---------------------------------------------------------------
    INPUTS: y: N-by-1 dependent variable
            X: N-by-K independent variable
---------------------------------------------------------------
    OUTPUT: Beta : OLS coefficient vector
            se : standard error of beta
---------------------------------------------------------------
```

Pretty handy! The precise structure of the help file is not important—you do not need to lay out your own help file as we have above. However, a clear statement of the purpose of the function, and of the dimension and description of the inputs and outputs, will be of huge help when you return to your code in the days, months, and years after writing it. Believe us— you will not remember what *that* X vector refers to by next week so make sure that your code reminds you of this!

(iii) The function assigns values to the matrices `Beta` and `se`. These are the names of the outputs in the first line. This means that when the function finishes running, it will return as outputs these assigned values.

(iv) Note that the function generally 'looks nice'.[10] In particular, note that there are subheadings to show the main parts of the calculation, comments (after the '%' symbol) to explain the operation of several of the lines of code, and the '=' signs are tabbed to the same alignment.

Let's use our function to repeat the regression from Section 1.1. Assuming that the 'auto' data from Section 1.1 is still in memory (which, remember, can be easily checked using the command `who`), we need simply pass this data to the function using the syntax of `OLS` that we have defined. We can do this from the command line:[11]

```
>> OLS(y, X)

ans =

    -0.0001
    -0.0058
    39.4397
```

As explained earlier, we do *not* need to refer to our variable names as `y` and `X`; these are the names that MATLAB will use *within* the function `OLS`, but this does not constrain the way that we *use* that function. For example, let's create two new variables, taking the values of `y` and `X`, and run the regression again:

```
>> barack = y;
>> hilary = X;
>> OLS(barack, hilary)
```

---

[10] Even if we say so ourselves . . .

[11] If you do not save the function in MATLAB's current working directory (which we can see using the `pwd` command), you will need to tell MATLAB where the M-file can be found. This can be done by using the `addpath` command. For example, if you have saved the function in a folder called 'C:/MATLABcourse/', you should enter `addpath C:/MATLABcourse`. Doing this, you'll come across a nice time-saving feature of MATLAB: tab completion. If you enter part of the path and press the $\boxed{\text{tab}}$ key, MATLAB will complete the path address if only one unique ending exists, or list all available ways the path could end if multiple endings are possible.

```
ans =

    -0.0001
    -0.0058
    39.4397
```

This is fine if we just want to display our regression results on the screen—but what if we want to store the results in a variable (say, OLS_Beta)? This is straightforward: we simply assign the variable as in Section 1.1, but have the variable refer to a calculation using our function:

```
>> OLS_Beta = OLS(barack, hilary)

OLS_Beta =

    -0.0001
    -0.0058
    39.4397
```

As expected, the function OLS returns the same regression results as in Section 1.1.

Comparing the output of our regression to the definition of the function OLS, you might ask the question: whatever happened to the variable se? When we programmed the file OLS.m, we specified the output as '[Beta, se]'—but, so far, OLS has reported only OLS_Beta. The reason is that, when we ran OLS, we have only *asked* for OLS_Beta: if we call a function from the command line, or assign the result of a function to a single variable, MATLAB will only return the *first* output variable. We can recover OLS_Beta *and* se by assigning *both* of these variables jointly:

```
>> [OLS_Beta, OLS_se] = OLS(barack, hilary)

OLS_Beta =

    -0.0001
    -0.0058
    39.4397

OLS_se =

    0.0002
    0.0006
    1.6216
```

We might also want to ask MATLAB to report a horizontal concatenation of OLS_Beta and OLS_se, just to make things look nice:

```
>> [OLS_Beta, OLS_se]

ans =

   -0.0001    0.0002
   -0.0058    0.0006
   39.4397    1.6216
```

Hopefully, we can now start to appreciate the beauty and simplicity of functions. *Yes*, it is true that functions can save us from repeating a lot of unnecessary typing—and, *yes*, it is true that functions can help to avoid careless mistakes. But the true beauty of functional programming is that we can replace a *number* with the *solution to a mathematical expression*—and do so with a syntax that is both simple and intuitive. For this reason, functions will be fundamental to everything we do in the rest of this book.

If you are ready for a breather from reading and want to write a function yourself, great! We once again point your attention to the exercises at the end of the chapter. Exercise (ii) asks you to write a post-estimation command.

1.3 A Simple Utility Function

A particularly important building block of microeconomic theory is the utility function, which maps the quantity of goods that an agent consumes to their payoff. Needless to say, this has all the ingredients to be used in a MATLAB function: it accepts inputs (goods consumed), it returns outputs (utility), and it is self-contained, depending entirely upon the inputs and a number of technology parameters.

Let's consider the Cobb-Douglas utility function. In this case, our output will be utility, u, and our inputs will be good 1, x_1, and good 2, x_2. For now, we will take a very simple form of the Cobb-Douglas function:

$$u(x_1, x_2) = x_1^{1/2} \cdot x_2^{1/2}. \tag{1.4}$$

Let's have a look at the function UtilitySimple to see how this would be set up in MATLAB.

UtilitySimple.m

```
1  function u = UtilitySimple(x1, x2)
2  % - - - - - - - - - - - - - - - - - - - - - - - - - - - - - - - - - -
3  % PURPOSE: calculate utility: 2 good Cobb-Douglas
4  %                specification
5  % - - - - - - - - - - - - - - - - - - - - - - - - - - - - - - - - - -
6  % USAGE: u : UtilitySimple(x1, x2)
7  % where: x1 : quantity of q1
8  %          x2 : quantity of q2
9  % - - - - - - - - - - - - - - - - - - - - - - - - - - - - - - - - - -
10 % OUTPUT: u : overall utility
11 % - - - - - - - - - - - - - - - - - - - - - - - - - - - - - - - - - -
12
13 u    =    (x1^0.5) * (x2^0.5);
14
15 return
```

Let's check that this works. We can check, for example, that the function returns correct answers for a few different bundles . . .

```
>> UtilitySimple(1, 4)

ans =

     2


>> UtilitySimple(3, 3)

ans =

    3.0000
```

Of course, we rarely want to use MATLAB merely to calculate a *single* number. We need an elegant way of dealing with multiple possible combinations of x_1 and x_2. Suppose that, for some reason, we want to find utility for $x_1 = 5$ and $x_2 \in \{1, \ldots, 10\}$. Create a vector x1 and a scalar x2 to represent this:

```
x1   = [1:10]';
x2   = 5;
```

We now have ten combinations of (x_1, x_2) for which we need to find $u(x_1, x_2)$. We *could* have UtilitySimple operate ten *separate* times—for example, using a loop.[12] But this is very inefficient. Instead, we should have

---

[12] We will introduce loops in Chapter 2.

UtilitySimple run *once*, and operate on the entire matrix x1. This is known as *vectorizing*. Having defined x1 and x2, we should be able simply to enter:

```
>> UtilitySimple(x1, x2)
Error using ^
Inputs must be a scalar and a square matrix.
To compute elementwise POWER, use POWER (.^) instead.

Error in UtilitySimple (line 13)
     u = (x1^0.5)*(x2^0.5);
```

But we have a problem! Look again at the function UtilitySimple. As it is written now, the function works perfectly well for *scalars* x1 and x2, but it does not work for vectors (or, more generally, for matrices). This is because the operators used there—the power operator and the multiplication operator—are understood by MATLAB to refer to matrices. We managed to get the correct answers when entering two scalars (for example, when we calculated UtilitySimple(1, 4)), because the scalar/matrix distinction did not matter in this simple case. But our function does not work for the more general case.

Fortunately, MATLAB has an elegant solution: we can modify both the power operator and the multiplication operator so they work 'element-by-element'. For both the power operator and the multiplication operator, we can do this by introducing a leading '.'. Let's go back and fix our function (which we will now just call Utility) to allow for this . . .

Utility.m

```
1   function u = Utility(x1, x2)
2   %----------------------------------------------------
3   % PURPOSE: calculate utility: 2-good Cobb-Douglas
4   %                specification
5   %----------------------------------------------------
6   % USAGE: u : Utility(x1, x2)
7   % where: x1 : quantity of q1
8   %              x2 : quantity of q2
9   %----------------------------------------------------
10  % OUTPUT: u : overall utility
11  %----------------------------------------------------
12
13  u    =    (x1.^0.5) .* (x2.^0.5);
14
15  return
```

We now have a utility function that is correctly defined for matrices. This is very powerful—among other advantages, we can now visualize our function very efficiently. Let's suppose that we want to see how our function behaves for $(x_1, x_2) \in [0, 3] \times [0, 3]$. We can create a *meshgrid* to cover this two dimensional space (discretized on unit intervals):

```
>> [x1, x2] = meshgrid([0:3], [0:3])

x1 =

         0     1     2     3
         0     1     2     3
         0     1     2     3
         0     1     2     3


x2 =

         0     0     0     0
         1     1     1     1
         2     2     2     2
         3     3     3     3
```

Hopefully, it is clear what is going on here: we have defined a matrix x1 and a matrix x2 such that x1 and x2 cover the grid $\{0, 1, 2, 3\} \times \{0, 1, 2, 3\}$. With a single operation, we can now calculate utility for this entire grid:

```
>> u        = Utility(x1, x2)

u =

         0          0          0          0
         0     1.0000     1.4142     1.7321
         0     1.4142     2.0000     2.4495
         0     1.7321     2.4495     3.0000
```

We can then visualize this with the `surfc` command:

```
>> surfc(x1, x2, u)
```

Of course, we would really like to visualize this over a finer grid. This is easy using `meshgrid`:

```
[x1, x2]   = meshgrid([0:.1:3], [0:.1:3]);
u          = Utility(x1, x2);
surfc(x1, x2, u)
```

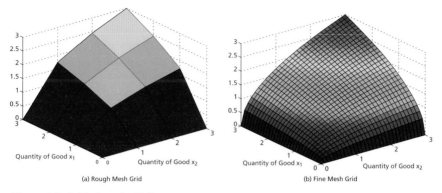

(a) Rough Mesh Grid   (b) Fine Mesh Grid

Figure 1.1 Cobb-Douglas Utility

As you will see in Figures 1.1a and 1.1b, the difference is quite stark, although the complexity in coding each example is virtually identical once we have set up our function. We hope that this is something which holds throughout much of this book: while some things may seem initially quite simple, the basic methods presented here can be generalized to solve and visualize functions of arbitrary complexity. We will consider a more complicated function in the exercises that follow, particularly Exercise (iii).

1.4 **Review and Exercises**

Table 1.1 Chapter 1 Commands

Command	Brief description
whos	Describes variables currently in memory
ans	Return the last item in memory
help	Describes a function along with its syntax
lookfor	Searches all M-files (including help files) for a keyword
ones	Creates an array of ones
cd	Changes current working directory
ls	Lists content of current working directory
pwd	Prints the location of the current working directory
dlmread	Reads in a comma-separated values file
disp	Prints text to the output window
inv	Inverts a matrix
size	Displays the size of an array
format	Sets the format of numerical output
mldivide	An efficient way to solve matrix division
function	Define a function which can be called from the command line
diag	Create vector of the diagonal elements of a matrix
addpath	Adds a directory to the places MATLAB searches when a command is called
meshgrid	Replicates vectors to form a rectangular matrix
surfc	Draws a three-dimensional surface plot

In this chapter, we have introduced MATLAB, its syntax, and a few useful examples that we suspect you have come across before (probably many
times). For now, we will hold off from suggesting any microeconometrics
readings. This will come in the chapters ahead! We do, however, suggest that
you experiment with MATLAB by writing your own scripts and functions. If
you come across issues that you do not yet know how to resolve, don't worry—
there will be many chances for further review in the pages ahead.[13] If you have
simple examples of problems from economics courses or ongoing research, feel
free to try to write simple codes to do these in MATLAB—or alternatively, for
more focused exercises, we provide questions below.

EXERCISES

(i) We have used MATLAB to recreate Stata's point estimates in a regression function.
Can you now generate the same standard errors? (*A useful hint*: remember that
$\widehat{\text{Var}}(\hat{\beta}) = (\boldsymbol{X'X})^{-1}\hat{\sigma}^2$, where $\hat{\sigma}^2 = (N-k)^{-1} \cdot (\boldsymbol{y} - \boldsymbol{X}\hat{\beta})' \cdot (\boldsymbol{y} - \boldsymbol{X}\hat{\beta})$, and k is the number of regressors (*including* the constant term).)

(ii) The OLS function introduced in this chapter returns point estimates and standard errors
for our regression coefficients. Try to write a function that can be used after OLS to
calculate a 95% confidence interval, and perhaps also t-tests and associated p-values.
(*A useful hint*: MATLAB's functions `tinv` and `tcdf` may be useful here.)

(iii) Suppose that we want to generalize our utility function slightly, so that we have:

$$u(x_1, x_2, \alpha) = x_1^\alpha \cdot x_2^{(1-\alpha)}.$$

Create a new function, `Utility2.m`, to accommodate this. Check that `Utility2.m`
matches the behaviour of `Utility.m` for the special case $\alpha = 0.5$. Repeat the visualization exercise. How does variation in α change the shape of u?

(iv) Suppose that a consumer has a utility function of the form:

$$u(x) = -\exp(-r \cdot x).$$

Also, suppose that x is drawn from a Normal distribution with mean μ and variance σ^2
(that is, $\mu \sim \mathcal{N}\left(\mu, \sigma^2\right)$). An insurance company offers the consumer a product with a
guaranteed lower limit, g. In effect, the insurance company says, '*If you buy our product
and $x < g$, we will pay the difference, so you will get g. If you buy our product and
$x > g$, we will do nothing, so you will just keep x.*'

(a) Interpret the parameter r.
(b) What is the certainty equivalent if there is no insurance product?

To answer this question, you may rely on the following result (which applies for this
special 'exponential-normal' case):

$$\mathbb{E}(u) = -\exp\left[\frac{1}{2} \cdot \sigma^2 \cdot r^2 - \mu \cdot r\right].$$

---

[13] Or even better, feel free to search around on the web! Along with MATLAB's own resources,
websites like stackoverflow.com are also likely to be of great use for you here.

(c) What is the expected utility with the insurance product? (Assume that the insurance company provides the product for free.)

To answer this question, you may rely on the following result (which, again, applies just for this special 'exponential-normal' case):

$$\mathbb{E}\left(u \,|\, x > g\right) = -\exp\left[\frac{1}{2} \cdot \sigma^2 \cdot r^2 - \mu \cdot r\right] \cdot \left[\frac{1 - \Phi(a + \sigma \cdot r)}{1 - \Phi(a)}\right],$$

where $a = \dfrac{g - \mu}{\sigma}$, and $\Phi(\cdot)$ is the *cdf* of the Normal.

(d) (For MATLAB . . .) Assume now that $\mu = 0$ and $\sigma^2 = 1$. Define $s(g, r)$ as the consumer's surplus—in utility terms—from having the insurance product. Graph the function $s(g, r)$ for $(g, r) \in (-3, 3) \times (0, 1.5)$.

2 The Agent Optimizes

> Nothing can have value without being an object of utility.
>
> Marx[*]

Now the fun really begins. Constrained optimization is a fundamental tool in economics. Essentially, there are two ways that optimization matters:

(i) We treat *agents* as optimizing—for example, we model consumers as maximizing utility, firms as maximizing profits, and so on.

(ii) We need to use optimization *ourselves* to find the best possible fit for our model (where 'best' is defined by some particular objective function).

In many respects, these two concepts are fundamentally different. On the one hand, we treat agents as if they optimize, as a way of *specifying* a model. On the other hand, we actually optimize, as a way of *estimating* the model. We should always keep these two concepts distinct in our minds. Yet each concept involves an optimization problem—and the principles and methods that we use for the two problems are remarkably similar.

In this chapter, we will consider the first of these cases—the optimizing agent. We will here use MATLAB to solve some common problems that we encounter in microeconomics.

2.1 Profit Maximization

Cast your mind back to your first term studying economics. It is highly likely that you were asked to solve a question similar to the following: *Given output prices, the cost of inputs and any capacity constraints, what levels of inputs should the firm select in order to maxmize profit?*

Before starting with our example, a brief word is needed on the structure of the profit maximization problem. Many of the optimization problems that we encounter in economics are formally called 'linear programming problems': they involve selecting the values of a set of choice variables to optimize some

---

[*] Marx K., *Capital* Volume 1, Progress Publishers, Moscow (1867).

linear objective function subject to a set of linear constraints. The general formulation of a linear programming problem is:

$$\min_{x} f'x \quad \text{such that} = \begin{cases} A \cdot x & \leq & b; \\ Aeq \cdot x & = & beq; \\ & lb \leq x \leq ub, \end{cases} \quad (2.1)$$

where x represents the vector of variables to be chosen, f, b, and beq are vectors, and A and Aeq are matrices of known coefficients. lb and ub denote the lower and upper bounds on the choice variable.

We will use the MATLAB function linprog to solve linear **programming** problems of this type. The basic linprog syntax is:

```
[x, fval, exitflag] = linprog(f, A, b, Aeq, beq, ...
                             lb, ub, x0)
```

The inputs f, A, b, Aeq, beq, lb, ub are defined as in Equation 2.1. x0 is an optional starting guess for the solution.

The outputs that we receive are: x (the optimal solution to the linear programming problem), fval (the value of the objective function at the optimal x), and exitflag (a variable that returns information about the optimization procedure, i.e. whether a minimum was reached, or whether the problem is infeasible or was not solved adequately).

Let's get our hands dirty and use linprog to solve a simple profit maxmization example. Suppose a farmer has 75 acres to plant with crops. She must decide how much to plant of crop a, wheat, and how much to plant of crop b, corn. The farmer operates in a perfectly competitive market where wheat commands a higher price than corn. The farmer's revenue function takes the form:

$$R(a, b) = 143a + 60b. \quad (2.2)$$

If the farmer were unconstrained, she would devote all the space to growing crop a. However, things are not so simple. Crop a requires more storage space than crop b and the number of storage units that the farmer has at her disposal is constrained to 4,000:

$$110a + 30b \leq 4000. \quad (2.3)$$

However, the seeds of crop b are more expensive and the farmer is credit constrained with only $\$15,000$ to spend on her initial outlay of crops:

$$120a + 210b \leq 15000. \quad (2.4)$$

Let's use MATLAB to answer the following:

(i) How much crop a should she plant?
(ii) How much crop b should she plant?
(iii) What is her optimal revenue?

The vector of choice variables, x, is structured as:

$$x = \begin{bmatrix} a \\ b \end{bmatrix} \tag{2.5}$$

Turning to the objective function, the farmer wants to maximize $R(a, b)$. However, linprog works to *minimize* the objective function. f, therefore, takes the form:

```
>> f = [-143; -60];
```

There are three inequality constraints that must also be respected: Equations 2.2, 2.3, and 2.4. These are captured by the A matrix and b vector:

```
>> A = [1, 1;110, 30;120, 210]

A =

     1      1
   110     30
   120    210

>> b = [75; 4000; 15000]

b =

      75
    4000
   15000
```

The final constraint is the lower bound—the farmer cannot plant a negative acreage of crops. There is no upper bound. Thus,

```
>> lb = zeros(2,1);
```

We now have all the elements of the linear program to pass to linprog as follows....

```
[crops, obj, exitflag] = linprog(f, A, b, [],...
                         [], lb)
```

What we have done here is call linprog with our linear objective function and inequality constraints, followed finally by the lower bound on the acreage of each crop. You may be wondering why we have included the two empty matrices [] in our code. In this problem, we do not have any strict equality constraints to worry about, but we do want to define a lower bound on the choice variables. Therefore, we have to declare the equality constraint arrays as empty.

Having passed the problem to `linprog`, it returns the optimal choice of crop a and b in the result vector `crops`, the value of the objective function in `obj` and the status of the optimization problem in `exitflag`.

```
>> [crops, obj, exitflag] = linprog(f, A, b, [],...
                              [],lb)
Optimization terminated.

crops =

    21.8750
    53.1250

obj =

  -6.3156e+03

exitflag =

      1
```

The value of the objective function (`obj`) needs relatively little explanation, beyond pointing out that, as expected, it is the negative of $R(a, b)$. The value of `exitflag` should be studied carefully whenever you optimize in MATLAB as it tells us why the optimization routine terminated. The MATLAB's help file for `linprog` lists all possible exit flags. An exitflag of 1 is returned when the function has converged to a solution. Weakly negative values of the exitflag correspond to infeasible problems or those that have terminated before the function has converged.

2.2 Utility Maximization

Many of the optimization problems that economists want to solve are not linear. For example, the utility maximizing consumer that we met in Chapter 1 chose a bundle of goods to maximize her *non-linear* Cobb Douglas utility function. Let's return to this example to introduce a central element of MATLAB's optimization routines: the `fmincon` command. This is a very powerful function for **min**imization subject to **con**straints, which you will use frequently in your MATLAB programming.

The fmincon command is used to solve optimization problems with the structure:

$$\min_{\mathbf{x}} f(\mathbf{x}) \quad \text{such that} = \begin{cases} \mathbf{A} \cdot \mathbf{x} \le \mathbf{b}; \\ \mathbf{Aeq} \cdot \mathbf{x} = \mathbf{beq}; \\ c(\mathbf{x}) \le 0; \\ ceq(\mathbf{x}) = 0; \\ \mathbf{lb} \le \mathbf{x} \le \mathbf{ub} \end{cases} \tag{2.6}$$

This is more general than a linear programming problem: the objective function $f(\mathbf{x})$ is not necessarily linear and we now allow for the non-linear constraints $c(\mathbf{x})$ and $ceq(\mathbf{x})$.

The basic fmincon syntax is:

```
[x, fval, exitflag] = fmincon(fun, x0, A, b, Aeq,...
                      beq, lb, ub, nonlcon, options)
```

Unlike linprog, fmincon requires us to specify an initial guess x0 from which it attempts to find the vector of choice variables x that minimizes the objective function fun subject to the user-specified constraints. nonlcon is a function that returns the value of $c(\mathbf{x})$ and $ceq(\mathbf{x})$.

Let's put fmincon to work by returning to our utility maximization example. As in the previous chapter, we will assume that the consumer's utility function is $u(x_1, x_2) = x_1^{1/2} x_2^{1/2}$, so that her maximization problem is:

$$\max_{x_1, x_2} x_1^{1/2} x_2^{1/2} \quad \text{subject to} \quad I = p_1 x_1 + p_2 x_2. \tag{2.7}$$

We have already defined this function in Utility.m. Before continuing, we must make two slight amendments to Utility.m so that it is easy to use with fmincon. fmincon works by adjusting a single vector of variables x. However, Utility.m is currently designed to take two arguments, the quantity of each good consumed. We amend the function to accept a single vector of quantities that we then simply 'unpack' into its individual elements, as you can see in the first two lines of the new function below. On the last line of the new function we take the *negative* of utility—given that, like linprog, fmincon works to minimize the objective function.

CobbDouglas.m

```
1  function u = CobbDouglas(X)
2  % - - - - - - - - - - - - - - - - - - - - - - - - - - - - - - - - - - -
3  % PURPOSE: calculate utility: 2 good Cobb-Douglas
4  %          specification
5  % - - - - - - - - - - - - - - - - - - - - - - - - - - - - - - - - - - -
6  % USAGE: u : CobbDouglas(X)
```

```
7   % where:  X : 2-by-1 quantity vector
8   %-------------------------------------------
9   % OUTPUT: u : overall utility
10  %-------------------------------------------
11
12  x1   =   X(1);
13  x2   =   X(2);
14  u    =   -(x1^0.5)*(x2^0.5);
15
16  return
```

Now that we have the utility function in the correct form, we can go about the business of building the budget constraint and optimizing! Let's imagine that our consumer has a total income of $100, faces goods prices of $4 and $7, and our initial guess is that she would consume 15 and 5 units respectively.

```
>> I   = 100;
>> P   = [4,7];
>> G   = [15,5];
>> lb  = [0,0];
```

Here we have introduced the right side of our budget constraint (the prices) as P, the left side as I, our initial guess as G, and a lower bound for the consumption of each good as lb.[1] This all comes together with the fmincon command as follows:

```
>> [consumption, u, exitflag] =...
                     fmincon(@CobbDouglas,...
                     G,P,I,[],[],lb)

consumption =

    12.5024    7.1415


u =

    -9.4491
```

---

[1] If we were intersted in checking whether our initial values satisfy the constraint we could check if $P'G < I$. In MATLAB:

```
» P'*G<I
ans =
1
```

implying that it *does* satisfy the constraint.

```
exitflag =

     5
```

What we have done here is call `fmincon` with our constraints and initial guess as parameters.[2] The only thing that might look a bit exotic is the @ symbol—this tells `fmincon` that `CobbDouglas` is a function. `fmincon` then returns the optimal consumption bundle: approximately 12.50 units of x_1 and 7.14 units of x_2.

So what does it mean to say that `exitflag = 5`? This message is mysterious but important. As we are now dealing with a non-linear objective function (and in other applications, possibly non-linear constraints), `fmincon` can only guarantee at best that a *local minimum* has been found. An exitflag of 5 is returned when the derivative of the objective function with respect to the choice variables is arbitrarily small. It is important to note that we can never be entirely sure that this point corresponds to a local minimum. Perhaps if we were to search over a finer set of points, `fmincon` would be able to find a smaller value for the objective function. Similarly, it may help `fmincon` if we were to provide an analytical expression for the Jacobian.[3]

Fortunately, MATLAB allows us great control over *how* we actually optimize—giving us the opportunity to check the accuracy of the solution. For example, we can ask MATLAB to use a particular algorithm, to set the value that the gradient must take at the 'optimum', to control the step size of the search path, and so forth. All of this is controlled the `optimset` command. Typing `optimset` at the command line lists many options, and the default values MATLAB assumes when using `fmincon`.

To illustrate, let's request that `fmincon` optimizes using the sequential quadratic programming (SQP) algorithm, rather than the default interior-point algorithm. In order to do this we define `opts` that contains our specific optimization specifications. This is passed to `fmincon` as the tenth argument, leaving another empty matrix in the ninth position because we do not have any non-linear constraints.

```
>> opts = optimset('algorithm', 'sqp');
>> [consumption,u,exitflag] =...
                    fmincon(@CobbDouglas,...
                    G,P,I,[],[],lb,[],[],opts)
```

---

[2] Note that here we treat the income constraint as non-binding, although generally it always will. An individual *could* choose to spend less than all their income, although the form of the Cobb-Douglas utility function implies that maximization of utility is achieved by spending all income on consumption.

[3] This often requires great cost to the programmer—both in terms of time and, sometimes, emotion....

```
consumption =

   12.5000    7.1429

u =

  -9.4491

exitflag =

      1
```

This change in optimization algorithm results in a small shift in our consumption values, and `fmincon` now returns an exitflag of 1, indicating that a local minimum has successfully been found. Usually, 'exitflag = 1' (or 'exitflag = 2') is a reasonable sign of success—though, as the previous discussion indicates, deciding on convergence criteria is often more a matter of art than science.

2.3 Simulating Economic Models

At its core, microeconometrics is about heterogeneity. If everyone faced the same set of choices, with the same constraints, and had the same preferences, then we would all behave in the same way—and there would be little advantage to collecting any dataset larger than $N = 1$. In many standard econometric models, we allow heterogeneity to enter through an additively separable 'error' term. Of course, there is nothing inherently wrong with this approach—but we may want to explore the consequences of introducing random variation in other parts of a model. We end this chapter with a simple illustration of how we might do this in MATLAB.

MATLAB offers us a wide range of tools for drawing random numbers from the specific distributions which are likely to underlie our Monte Carlo Simulations. Table 2.1 lists a number of these distributions, and their associated MATLAB commands.

Table 2.1 Random Number Generators

Distribution	Command
Uniform	`rand()`
Normal	`randn()`
Lognormal	`lognrnd()`
Multivariate Normal	`mvnrnd()`
and many others...	`random(distbn,)`

It is sometimes argued that different economic actors face different prices, based on their individual characteristics. (For example, information asymmetries may lead different firms to face different factor costs.) Suppose, then, that we maintain our earlier model of a consumer with Cobb-Douglas utility. However, now imagine that, across the population, consumers face uniform variation in the price of good 1:

$$p_1 \sim U(50, 100). \qquad (2.8)$$

Suppose that we want to simulate the resulting distribution of consumption bundles. Given the tools that we have discussed, this is easy.

The following script gives a simple illustration of how this can be done—we simulate the choices of 100 individuals who have identical preferences but face different prices. In line 9, we generate 100 shocks to p_1 by drawing errors from the random uniform distribution and multiplying by 50. Once these errors are added to P, we will have $p_1 \sim U(50, 100)$, as required. In part 2 of the code, the starting value, options, and lower bound on the optimization procedure are set. We 'preallocate' a results vector c that is first made up of elements of 'not a number', NaN. We will use this vector to store optimized consumption for each individual.

On line 21, we start a for loop. This command allows us to loop over the integers from 1 to 100 (as reps = 100) setting i to each value in turn. For each value of i, the lines of code 20–23 are executed. First, the price for individual *i* is calculated and, given this, the upper bound on consumption is set. On line 22, this is passed to fmincon and CobbDouglas is maximized subject to the individual-specific budget constraint. Finally, we use the function cdfplot to draw the empirical CDF of the consumption of good 1 as shown in Figure 2.1.

ConsumptionSim.m

```
1  %----(1) Setup, simulation of random variation  --
2  clear
3
4  % random variation in prices
5  P           = [50, 200];
6  reps        = 100;
7  pshock      = [rand(reps, 1) * 50, zeros(reps, 1)];
8
9  I           = 10000; % income
10
11 %--- (2) Determine optimal consumption in
12 %           each case ----
13 x0          = [1, 1];
14 lb          = [0, 0];
```

```
15  opts         = optimset('algorithm', 'sqp',...
16                  'display', 'off');
17
18  C            = NaN(reps, 2);
19  for i = 1:reps
20    TempP      = P + pshock(i, :);
21    ub         = I./TempP;
22    c(i, :) = fmincon(@CobbDouglas,[1, 1],TempP,...
23               I,[],[],lb,ub, [], opts);
24  end
25
26  %---(3) Visualize results ----------------------
27  subplot(1, 2, 1)
28  scatter(c(:,1), c(:,2))
29  xlabel('Good 1 Consumption')
30  ylabel('Good 2 Consumption')
31
32  subplot(1,2,2)
33  cdfplot(c(:, 1))
34  xlabel('Good 1 Consumption')
35  ylabel('F(p_1)')
```

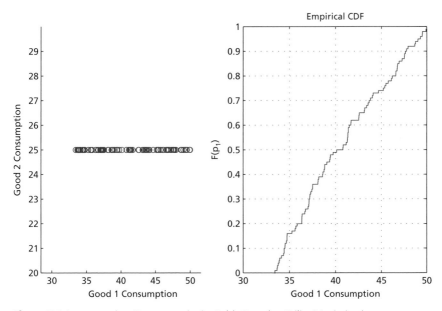

Figure 2.1 Incorporating Heterogeneity in Cobb-Douglas Utility Maximization

2.4 **Review and Exercises**

Table 2.2 Chapter 2 Commands

Command	Brief description
axis	Set minimum and maximum for graph axes
cdfplot	Draws the empirical CDF of a vector
disp	Print contents of an array
fmincon	Routine to minimize an objective function subject to linear or non-linear constraints
for	Repeats a command or series of commands a specified number of times
hold on	Keep current plot in graph window, and add another plot to the output
if	Evalutes an expression and executes a group of statements if it is true
Inf	Create an array of infinite values
linprog	Routine to minimize a linear objective function subject to linear constraints
num2str	Convert number to string
optimset	Provides control over the optimization process in MATLAB
plot	Draw a two-dimensional graph
rand	Allows for a (pseudo-)random draw from a uniform(0,1) distribution
randn	Allows for a (pseudo-)random draw from a normal distribution
rng	Sets MATLAB's pseudo random number generator at a replicable point
subplot	Allows for multiple graphs on the same plot window
scatter	Bivariate scatter plot
title	Add a title to the plot window
xlabel	Label x-axis in the plot window (allows for LATEX style parsing)

EXERCISES

(i) A fisherwoman must decide how many salmon and how many trout to catch. She can sell a salmon for $12 but can only sell a trout for $7. However, salmon requires twice as much storage space as trout and she only has room for 1,000 trout on her boat. Further, there is a quota limiting the numbers of fish she can catch. She may only catch 1,600 tokens worth of fish, where each salmon is worth 3 tokens and a trout only 2 tokens. Use linear programming to determine how many salmon and how many trout the fisherwoman should catch. How does the solution vary with the price of trout?

(ii) Using the function CobbDouglas, simulate variation in income. Use the simulated results to plot Engel curves for x_1 and x_2.

(iii) Using the function CobbDouglas, suppose that consumers with higher income *also* tend to face lower costs for good x_1. Show how a simulation method could be used to think about consumption bundles in this case. (*Hint*: mvnrnd may be useful . . .)

3 The Economist Optimizes

> We balance probabilities and choose the most likely. It is the scientific use
> of the imagination.
>
> Sherlock Holmes, *The Hound of the Baskervilles*[*]

Economists are human beings too. In the previous chapter, we considered
consumers as optimizing their choices to maximize utility. In this chapter, we
discuss ways that we, as economists, can choose parameters to optimize the
fit of our models. Specifically, we will use MATLAB's constrained optimization
techniques to estimate model parameters by Maximum Likelihood and by the
Generalized Method of Moments.

3.1 Maximum Likelihood

In Chapter 1, we ran an OLS regression by inverting a matrix. In this chapter,
we introduce Maximum Likelihood in the same simple setting. Famously, the
log-likelihood associated with the classical regression model (Equation 1.1)
looks like this:

$$\ell(\boldsymbol{\beta}, \sigma^2; \boldsymbol{y} \mid \boldsymbol{x}) = -\left(\frac{N}{2}\right) \ln 2\pi - \left(\frac{N}{2}\right) \ln \sigma^2 - \left(\frac{1}{2\sigma^2}\right) (\boldsymbol{y} - \boldsymbol{\beta} \boldsymbol{x})'(\boldsymbol{y} - \boldsymbol{\beta} \boldsymbol{x}).$$

$$(3.1)$$

The set of Maximum Likelihood parameters is then the $(\boldsymbol{\beta}', \sigma^2)'$ vector that
maximizes the log-likelihood function. Following Chapter 2, you will remem-
ber that we can combine MATLAB's optimization routines with our own functions. We can use `fmincon` to maximize ℓ by searching over combinations of
the parameters $\boldsymbol{\beta}$ and σ^2.

As always, the maximization process begins by converting our objective
function into a MATLAB function. As we alluded to earlier, it is necessary to
optimize over a *single* vector rather than over two separate objects when using
`fmincon`—so we define $\boldsymbol{\theta} = (\boldsymbol{\beta}', \sigma^2)'$, and optimize over $\boldsymbol{\theta}$. The function
`NormalML` defines the log likelihood in this way (multiplied by -1 to be
suitable for minimization by `fmincon`).

---

[*] Conan Doyle, A., *The Hound of the Baskervilles*, George Newnes (1902).

NormalML.m

```matlab
 1  function LL = NormalML(theta,y,x)
 2  %-----------------------------------------------
 3  % PURPOSE: calculates the likelihood function
 4  % given an unobserved stochastic error term
 5  % which is distributed according to a normal
 6  % distribution.
 7  %-----------------------------------------------
 8  % USAGE: LL : NormalML(theta,y,x)
 9  % where: theta : parameter vector [beta;sigma]
10  %             y : N-by-1 dependent variable
11  %             x : N-by-K independent variable
12  %-----------------------------------------------
13  % OUTPUT: LL = log-likelihood value
14  %-----------------------------------------------
15
16  %----- (1) Unpack stats -----------------------
17  N     =    length(y);
18  K     =    size(x,2);
19
20  Beta  =    theta(1:K);
21  sig   =    theta(K+1);
22
23  u     =    y - x*Beta';
24
25  %----- (2) Likelihood function ----------------
26  LL    =    -(N/2)*log(2*pi)-(N/2)*log(sig^2)-...
27             (1/(2*sig^2))*(u'*u);
28  LL    =    -LL;
29
30  return
```

We can again use `fmincon` to maximize the log likelihood. This is a constrained optimization problem: $\sigma^2 > 0$. To see this in practice, let's return to the trusty `auto.csv` example of Chapter 1. To do this, we will return to our parameter estimates from Section 1.1. There we had defined a matrix X and a vector y which had come from the auto dataset. We will ask you to re-enter or recall these matrices (perhaps they are still in MATLAB's working memory, in which case you need do nothing), and we will use this data to see whether our likelihood function allows us to recover the regression estimates we calculated in Chapter 1.

In the case of this problem, there is no clear upper and lower bound that we necessarily want to impose on our parameters, but we will define some values to limit the domain over which MATLAB searches.[1] Here we estimate four parameters that correspond to the three $\hat{\beta}$'s, along with $\hat{\sigma}^2$. First, let's set up the following lower and upper bounds for each of these parameters:

```
>> lb  = [-1000, -1000, -1000, 0];
>> ub  = [1000, 1000, 1000, 100];
```

We will also define an initial range of values from which MATLAB should begin its search. Although we have quite a good idea of what these parameters should look like from our regression in Chapter 1, we will pretend this is not the case, and be reasonably agnostic with our initial choice:

```
>> theta0  = [0, 0, 0, 1];
```

Finally, we want to define a number of optimization options. So far we have only just scratched the surface of the optimset options for fmincon. Previously we had decided to use the sqp algorithm and then let MATLAB decide on all the other values. Below we define a few of more of these options (as a reminder, the full list of options can be seen by typing optimset in your MATLAB window).

```
>> opt = optimset('TolFun',1E-20,'TolX',1E-20,...
         'MaxFunEvals',1000, 'Algorithm', 'sqp');
```

Here we have allowed MATLAB to keep running until subsequent changes in the objective function and the optimands are very tiny (10^{-20}) and at the same time allowed it to evaluate the objective function a sufficient number of times to find an answer. Together these options should allow for a very precise solution for the values $\boldsymbol{\theta}$.

We have all our ingredients: we have defined an objective function, the constraints, the starting point, and the optimization options. With all of this, it is now just a matter of letting MATLAB get to work . . .

```
>> fmincon(@(theta)NormalML(theta,y,X), theta0,...
         [], [], [], [], lb, ub, [], opt);
```

In the last chapter, we used the @ symbol to show that CobbDouglas was a function. Here we use it in a slightly more complicated way—to declare which of the function inputs is our choice variable. In MATLAB parlance, this is known as the 'function handle'. The NormalML(theta,y,X) part of the argument tells fmincon to minimize the function NormalML given the data y and X

---

[1] MATLAB also has a function for unconstrained optimization: fminunc. But as we suggested above, our problem is constrained because we have $\sigma^2 > 0$. Further, even if our problem were not constrained by economic or econometric theory, it may still be useful to use fmincon and to impose some bound constraints on 'reasonable' values of the parameter space.

(which we defined in Chapter 1), while @(theta) tells fmincon that the choice variable is theta.

You should play around with this code until you feel comfortable with the various moving parts. For example, try varying the optimization settings, the lower and upper bound, and the starting point. Depending upon the settings you use, the output will look something like the following:

```
Local minimum possible. Constraints satisfied.

fmincon stopped because the predicted change in the
objective function is less than the selected value
of the function tolerance and constraints are
satisfied to within the default value of the
constraint tolerance.

No active inequalities.

ans =

   -0.0001    -0.0058    39.4396    3.3842
```

Importantly, we see that with these settings our ML estimator correctly finds the same OLS estimate as in Chapter 1. Of course, it was much easier to obtain these values by simply inverting a matrix, as we did in Chapter 1. For this reason, we would never actually use Maximum Likelihood to run OLS—but OLS is a great way to illustrate the practical challenges of ML in a simple context.

Before moving on from this section, it is worth pointing out that optimset provides a whole range of useful options. If, for example, we are interested in producing a graph of convergence of the objective function to its minimum value then we could take advantage of: 'PlotFcns','optimplotfval', while if we are interested in seeing a larger range of output including the procedure and the value of the objective function, we could specify 'Display','iter' as part of our optimset command. In Figure 3.1 we see the output from the PlotFcns option. We have taken the absolute value of the likelihood function so it appears to be converging on a positive value from above, although in reality it will be converging on a negative value from below.

3.2 **Generalized Method of Moments**

Maximum Likelihood is a workhorse of microeconometrics, but is certainly not the only nag in the stable. One of the most flexible ways of estimating

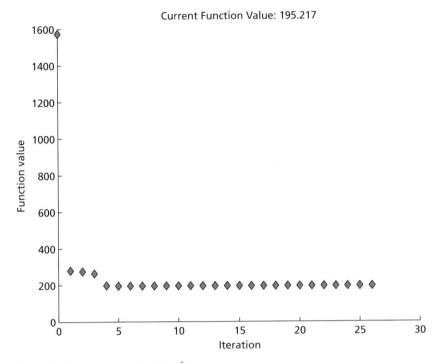

Figure 3.1 Convergence of $-\ell(\boldsymbol{\beta},\sigma^2;y\,|\,x)$

is by the Generalized Method of Moments (GMM). When using GMM, we define a number of population moments that are true in our model.[2] We then estimate the parameters of our model using the principle of analogy. This involves setting the identical sample moments to zero.

Like ML, GMM requires us to optimize an objective function, and again (for the last time, we promise!) the most simple example comes from estimating parameters in a linear regression model. You will remember from an early econometrics class that the basic Gauss-Markov assumptions imply that the following moments will hold in the population for the linear regression model:

$$\mathbb{E}[\boldsymbol{\varepsilon}|X] = 0. \tag{3.2}$$

From this, it follows that:

$$\mathbb{E}[X\boldsymbol{\varepsilon}] = \mathbb{E}[X(y - X\boldsymbol{\beta})] = 0. \tag{3.3}$$

---

[2] There are many excellent references if you are interested in further details on moment-based estimation. For example, Hall (2005), Cameron and Trivedi (2005), or for those particularly interested, the original Hansen (1982) article.

We can use Equation 3.3 to form the sample moment conditions. These sample moments are of the following form:

$$m = \frac{1}{N} \left[\sum_{i=1}^{N} X_i(y_i - X_i \beta) \right] = 0. \tag{3.4}$$

This is the moment vector and will be of dimension $1 \times K$, where K is the number of explanatory variables in our model.

The fundamental idea in GMM is that our estimates $\hat{\beta}$ should be those values that drive the weighted quadratic distance mWm' to zero (or as close to zero as possible if we have more moments than coefficients). For consistent estimation we simply need to ensure that our weight matrix W is positive semi-definite (such as an identity matrix). However, for efficiency reasons we may be interested in using other weight matrices such as those discussed by Hansen (1982). We return to precisely this point in Exercise (iii) at the end of this chapter.

Let's look at the function GMMObjective, which calculates the value of the moment equations. This function accepts as arguments our observed y and X data, and a proposed value for the vector of parameter estimates $\hat{\beta}$.

GMMObjective.m

```
1   function Q = GMMObjective(Beta,y,X)
2   %----------------------------------------------
3   % PURPOSE: calculates the moments of a linear
4   % regression model. The true method of moments
5   % estimate of beta occurs when Q=0.
6   %
7   % The series of moments that are being fitted
8   % here are: [E(X_1*u)=0 E(X_2*u)=0...E(X_k*u)=0]
9   %----------------------------------------------
10  % USAGE: Q : GMMObjective(Beta,y,X)
11  % where: Beta : parameter vector
12  %             y : N-by-1 dependent variable
13  %             X : N-by-K independent variable
14  %----------------------------------------------
15  % OUTPUT: Q : value of moment conditions
16  %----------------------------------------------
17
18  %----- (1) determine sample size N and number of
19  %             coefficients K
20  N = length(y);
21  K = size(X,2);
```

```
22
23   %----(2) Calculate u and generate the identity
24   %              weight matrix
25   u = y - X*Beta;
26   W = eye(K);
27
28   %----- (3) Generate moment vector ----------
29   m = 1/N*u'*X;
30   Q = m*W*m';
31
32   return
```

To calculate the estimates, all we now need to do is minimize the function GMMObjective. In this case, we will use fminsearch rather than fmincon because our optimization problem is unconstrained. Let's again return to the auto.csv file we worked with in Chapter 1. In the code below, we load this into MATLAB, and estimate $\hat{\beta}$ by GMM:

```
>> DataIn   = dlmread('auto.csv');
>> X        = [ones(74,1) DataIn(:,2:3)];
>> y        = DataIn(:,1);
>>[Beta,Q]  = fminsearch(@(B) GMMObjective(B,y,X),...
               [10,0,0]', optimset('TolX',1e-9));

Beta =

    39.4397
    -0.0001
    -0.0058


Q =

    6.3475e-20
```

By minimizing this objective function, MATLAB gets very close to $Q = 0$, and recovers the identical estimates that we found previously. You will notice that we have here specified some different optimization settings in our call to fminsearch. In this example, we have exactly as many moments as unknowns. This is called 'just identification'. Whenever GMM is just identified, we should obtain $Q = 0$—or something very close to this. In any situation where your model is just-identified and Q is not extremely close to zero, your optimization settings likely need fine tuning.

3.3 **Review and Exercises**

Table 3.1 Chapter 3 Commands

Command	Brief description
length	Determines the number of rows in a matrix
pi	The mathematical constant π
log	Return the natural logarithm of a number of matrix
fminunc	Routine to minimize an objective function with no constraints
fminsearch	Routine to minimize an objective function with no constraints
optimset	Provides control over the optimization process in MATLAB
eye	Creates an identifty matrix

In this chapter, our second on optimization, we have gone through some typical estimation methods that will become the base of integrating our theories with our data in later chapters. If you would like to read more on the econometric theory behind these estimation methods, we suggest that you turn to your favourite econometrics or statistics textbook, for example, Cameron and Trivedi (2005), Wooldridge (2010), or Casella and Berger (2002).

EXERCISES

(i) Write an alternative Maximum Likelihood estimator. Rather than an OLS estimator, try writing MATLAB code to estimate a probit model. Remember that, in this case, the log-likelihood function looks like: $\ell(\boldsymbol{\beta}; \mathbf{y}|\mathbf{x}) = \sum_{i=1}^{N} \{y_i \cdot \ln \Phi(\boldsymbol{\beta}x_i) + (1 - y_i) \cdot \ln[1 - \Phi(\boldsymbol{\beta}x_i)]\}$. Try this code using the auto dataset as before. Estimate 'probit foreign length weight' in Stata. Ensure that your code replicates these results in MATLAB.

(ii) Consider again the GMM and ML estimators discussed in this chapter. How could you calculate the standard errors of the above estimators in MATLAB? Do these agree with those we calculated in Chapter 1? Econometrically, is this what you would expect?

(iii) When we estimate our linear regression by GMM, we use the identity matrix (eye()) for W in our quadratic loss function. However, for efficiency reasons, we may prefer alternative weighting matrices. Perhaps the most popular option is two-step GMM, where the first step is as in our code, and in the second step we use an optimal weight matrix $\widehat{W}$ that is based on the first stage estimates $\hat{\boldsymbol{\beta}}_{(1)}$. Can you write your own MATLAB code for this estimator? As a reminder, the optimal weight matrix can be written as

$$\widehat{W} = \left[\frac{1}{N} \left(\sum_{i=1}^{N} X_i (y_i - X_i \hat{\boldsymbol{\beta}}_{(1)}) (y_i - X_i \hat{\boldsymbol{\beta}}_{(1)}) \right)' \right]^{-1}.$$

(iv) Some code in this chapter has estimated $\boldsymbol{\beta}$ using just-identified GMM. However, in this case, we could also consider estimating via traditional Method of Moments, which simply involves setting each moment exactly equal to zero (rather than minimizing the quadratic distance). How would you code a Method of Moments estimator for $\boldsymbol{\beta}$ in MATLAB? (*Hint:* The fsolve function may be useful in this case.)

Part II
Discrete Choice

4 Discrete Multinomial Choice

> But you must understand, sir, that a person is either with this court or he must be counted against it; there be no road between.
>
> Arthur Miller, *The Crucible**

In this chapter, we consider several models of discrete choice. This will be useful for at least two reasons. First, we often treat decision makers as choosing from a finite set of options. In some cases, this reflects the underlying economic reality—will you choose to travel by car, bus, or train? In other cases, it is because researchers sometimes find it useful to discretize continuous variables.

Second, discrete choice models are a useful way of introducing the idea of *nested optimization*. In Chapter 2, we discussed optimization by economic agents, while in Chapter 3 we discussed optimization for the purposes of estimation. Applied research often nests the former inside the latter. That is, we specify a model in which agents optimize, and then choose the structural parameters that best fit the data.

In this chapter, we will focus on a simple problem: the demand for various modes of transport. We will assume throughout that we have data on $N = 1{,}000$ individuals (indexed by i), and that, for each individual, we observe a single covariate, income (which we will denote x_i). This is a simple application of a problem famously studied by Daniel McFadden and others, who developed new methods in discrete choice theory to estimate consumer demand for the San Francisco Bay Area Rapid Transit system ('BART').

We will consider three different models for this problem, with increasing degrees of sophistication. First, we present a (binary) logit model. Second, we consider a multinomial outcome model, making a strong distributional assumption about the error terms (the multinomial logit). Finally, we discuss a multinomial model with a more flexible distributional assumption (the multinomial probit). The chapter closely follows the excellent discussion in Train (2009, Chapters 3 and 4).

---

* Miller, A., *The Crucible*, Viking Books (1952).

4.1 **Binary Logit**

Let's begin with a simple decision problem: should a commuter (i) drive to work, or (ii) take other modes of transport?[1] We can capture this choice using a binary outcome variable y_i, with $y_i = 1$ if the commuter drives to work.

We can model this problem as a binary logit. Assume that, if individual i chooses $y_i = 1$, she or he receives utility U_{i1}, with U_{i0} defined analogously. We specify U_{ij} as the sum of (i) a deterministic function of the data ($V_{ij}(x_i)$) and (ii) a random noise variable (ε_{ij}):[2]

$$U_{i0} = V_{i0}(x_i) + \varepsilon_{i0}; \tag{4.1}$$

$$U_{i1} = V_{i1}(x_i) + \varepsilon_{i1}. \tag{4.2}$$

The solution to this model is straightforward:

$$y_i = \begin{cases} 1 \text{ if } V_{i1}(x_i) + \varepsilon_{i1} \geq V_{i0}(x_i) + \varepsilon_{i0} \\ 0 \text{ otherwise.} \end{cases} \tag{4.3}$$

$V_{i0}(x_i)$ and $V_{i1}(x_i)$ could be specified very flexibly—for example, we might have a sophisticated structural model that relates income and the representative utility of driving to work. However, to keep things simple, we will assume that $V_{ij}(x_i)$ is linear in x_i: $V_{ij}(x_i) = \alpha_{0j} + \alpha_{1j} \cdot x_i$. Having assumed a linear form for $V_{ij}(x_i)$, the solution can be expressed as:

$$y_i = \begin{cases} 1 \text{ if } \beta_0 + \beta_1 \cdot x_i + \mu_i \geq 0 \\ 0 \text{ otherwise,} \end{cases} \tag{4.4}$$

where $\beta_0 \equiv \alpha_{01} - \alpha_{00}$, $\beta_1 \equiv \alpha_{11} - \alpha_{10}$ and $\mu_i \equiv \varepsilon_{i1} - \varepsilon_{i0}$.

To close the model, we must make some assumptions about the distribution of ε_{ij}. Critically, we assume that ε_{ij} is independently and identically distributed according to a Type I Extreme Value distribution:

$$\Pr(\varepsilon_{ij} \leq z \,|\, x_i) = \exp(-\exp(-z)). \tag{4.5}$$

That is, we assume that ε_{ij} is independent across individuals (*i*) *and* across options faced by a given individual (*j*). It follows from this assumption that μ_i has a logistic distribution:

$$\Pr(\mu_i \leq z \,|\, x_i) = \frac{\exp(z)}{1 + \exp(z)} \equiv \Lambda(z). \tag{4.6}$$

---

[1] We will treat these as mutually exclusive: if the commuter drives for *any part* of his or her journey, we will treat this as driving to work.

[2] Train (2009, p.15) calls $V_{ij}(x_i)$ the 'representative utility'.

This allows us to write the conditional probability of choosing to drive to work as:

$$\Pr\left(y_i = 1 \mid x_i\right) = \Pr\left(\beta_0 + \beta_1 \cdot x_i + \mu_i \geq 0\right) \quad (4.7)$$

$$= \Lambda\left(\beta_0 + \beta_1 \cdot x_i\right). \quad (4.8)$$

where Λ is the logistic *cdf*.

4.1.1 SIMULATING THE MODEL

When coding your own estimator—no matter how simple—you should *always* start by simulating your model. That is, you should generate a fake dataset and check that you can recover reasonable estimates of the true parameters. If you cannot estimate successfully using this fake data, you will *never* feel confident with real data.

Let's start, then, by simulating our model for the simple case $(\beta_0, \beta_1) = (0.5, 0.5)$. First, let's seed the random number generator (using `rng`), create a vector `Beta`, and generate a normally distributed x_i...

```
>> rng(1)
>> N        = 1000;
>> Beta     = [0.5, 0.5]';
>> income = randn(N, 1);
>> x        = [ones(N, 1), income];
```

The function `SimulateBinaryLogit` generates binary outcomes according to the logit specification; it outputs the outcome variable (`y`), and the simulated utility (`utility`). Pay special attention to the command `epsilon=-log(-log(rand(N, J)))` on line 17—how does this draw from a Type 1 Extreme Value distribution?

SimulateBinaryLogit.m

```
1  function [y,utility]=SimulateBinaryLogit(x,Beta)
2  %------------------------------------------------
3  % PURPOSE: generate binary outcomes according to
4  % the logit specification.
5  %------------------------------------------------
6  % INPUTS:   x : NxJ vector of independent variable
7  %           Beta : Jx1 parameter vector
8  %------------------------------------------------
9  % OUTPUT:   y : N-by-1 binary choice vector
10 %------------------------------------------------
```

```
11
12  N                    = size(x, 1);
13  J                    = length(Beta);
14
15  %----- (1) simulate values for epsilon &, based
16  %             on this, utility ----
17  epsilon              = -log(-log(rand(N, J)));
18  Beta_augmented       = [Beta, 0 * Beta];
19  utility              = x * Beta_augmented + epsilon;
20
21  %----- (2) simulate the choice for each
22  %             individual ------
23  [~, choice]          = max(utility, [], 2);
24  y                    = (choice == 1);
25
26  return
```

4.1.2 ESTIMATING THE MODEL

Now that we have some simulated data to play with, we can move on to estimation. The log-likelihood associated with the binary logit model is given as:

$$\ell_i\left(\beta_0, \beta_1; y_i \mid x_i\right) = y_i \cdot \ln \Lambda\left(\beta_0 + \beta_1 \cdot x_i\right) +$$

$$(1 - y_i) \cdot \ln\left[1 - \Lambda\left(\beta_0 + \beta_1 \cdot x_i\right)\right]. \tag{4.9}$$

The function BinaryLogitLL evaluates $-1 \times \ell_i\left(\beta_0, \beta_1; y_i \mid x_i\right)$ from Equation 4.9.

BinaryLogitLL.m.m

```
1   function [LL, ll_i] = BinaryLogitLL(Beta, y, x)
2
3   %-------------------------------------------------
4   % PURPOSE: Calculate the log-likelihood for the
5   %             binary logit
6   %-------------------------------------------------
7   % INPUTS: Beta : Kx1 parameter vector
8   %             y : Nx1 vector of dependent variable
9   %             x : NxK matrix of independent variables
10  %-------------------------------------------------
11  % OUTPUT: LL : scalar log likelihood
12  %             ll_i : log-likelihood contribution of
```

```
13   %                         observations
14   %- - - - - - - - - - - - - - - - - - - - - - - - - - - - - - - - - - - -
15
16   Lambda_xb    = exp(x * Beta)./(1 + exp(x * Beta));
17
18   ll_i         =             y  .* log(Lambda_xb)  ...
19                       + (1 - y)  .* log(1 - Lambda_xb);
20
21   LL           = -sum(ll_i);
22
23   return
```

We can use `BinaryLogitLL` and `fmincon` to find the Maximum
Likelihood parameters for the binary logit model . . . .

```
>> options     = optimset('Algorithm', 'sqp',...
                  'Display', 'iter');
>> Beta_init = [0; 0];
>> lb         = [-10; -10];
>> ub         = [10; 10];
>> [EstBetaML, LL, exitflag] =...
     fmincon(@(parameters)...
     BinaryLogitLL(parameters, y, x), Beta_init,...
     [], [], [], [], lb, ub, [], options)

EstBetaML =

     0.6041
     0.5739

LL =

   601.1168

exitflag =

     2
```

Hopefully, you succeed in recovering reasonably good estimates of $(\beta_0, \beta_1) =$
$(0.5, 0.5)$. You should flip back and forth between simulation and estimation,
checking the performance of the estimator. How does the estimator perform
for different values of (β_0, β_1)? How does it perform for different sample
sizes?

4.1.3 ESTIMATING BY SIMULATION

The binary logit is beautifully straightforward because we can write an analytical expression for the log-likelihood. From there, we need simply to code the log-likelihood and optimize. However, now suppose—purely for the sake of exposition—that for some reason we are unable to write Equation 4.9. That is, suppose that we *cannot* find an analytical expression for the log-likelihood. It would be a pity if this were to prevent us from estimating our model. If it did, we would be forced to change our model purely for the convenience of our code, which is never an attractive option.

Fortunately, all would not be lost. So far, we have *simulated* our model, and we have *estimated*. But we can do more—we can combine the two methods, to *estimate by simulation*. For now, this is purely a pedagogical exercise—but this idea is fundamental when we need to estimate more flexible models later in this chapter.

The idea is simple. We have a model that predicts how observable behaviour (in this case, the choice of y_i) depends upon an observable covariate (income, x_i) and a parameter vector ($\beta = (\beta_0, \beta_1)$). We are able to simulate this model for different values of β. We also have some real data. For some values of β, our simulated data will look vastly different to our real data—for example, if we choose $\beta_1 = 0$, we will simulate data in which y_i does not vary with x_i, which is unlikely to be true in our real data. For other values of β, our simulated data will look quite similar to our real data. We can estimate by choosing β so that our simulated data 'looks like' our real data.

There are many ways of doing this. Our model relies upon an assumption about the entire distribution of the unobservable (that is, we have a fully parametric model)—so we can estimate by Maximum Simulated Likelihood ('MSL'). The basic idea of MSL is simple: instead of calculating probabilities using an analytical expression, we should simulate our model multiple times, and use the sample probabilities that are generated from the simulated data.

Formally, suppose that we have R replications of a simulation algorithm. Then, for each respondent i, we will have a simulated series of binary outcomes, $\{\tilde{y}_{i1}, \ldots, \tilde{y}_{iR}\}$. For each respondent i, all of their simulated outcomes are generated using the same value of x_i—this is the sense in which we are simulating conditional on the values of x_i in the data. Then, for individual i, the simulated conditional probability of choosing $y_i = 1$ is simply:

$$\tilde{P}(x_i; \beta_0, \beta_1) \equiv \frac{1}{R} \cdot \sum_{r=1}^{R} \tilde{y}_{ir}. \tag{4.10}$$

For individual i, the simulated log-likelihood is therefore:

$$\tilde{\ell}_i \left(\beta_0, \beta_1; y_i \mid x_i \right) = y_i \cdot \ln \tilde{P}(x_i; \beta_0, \beta_1) + (1 - y_i) \cdot \ln \left[1 - \tilde{P}(x_i; \beta_0, \beta_1) \right]. \tag{4.11}$$

For given parameter values β_0, β_1,

$$\lim_{R \to \infty} \tilde{\ell}_i\left(\beta_0, \beta_1; y_i \mid x_i\right) = \ell_i\left(\beta_0, \beta_1; y_i \mid x_i\right). \qquad (4.12)$$

This is the basic idea behind Maximum Simulated Likelihood.

The function `BinaryLogitSimulatedLL(Beta, y, x, R)` calculates $-1 \times \tilde{\ell}_i\left(\beta_0, \beta_1; y_i \mid x_i\right)$. Notice how this code nests the function `SimulateBinaryLogit` —the very same function that we used earlier to generate our outcome variable.

BinaryLogitSimulatedLL.m

```
1  function [LL, ll_i] = ...
2  BinaryLogitSimulatedLL(Beta, y, x, R)
3  %------------------------------------------------
4  % PURPOSE: calculate simulated likelihood
5  %------------------------------------------------
6  % INPUTS: Beta : Kx1 parameter vector
7  %              y : Nx1 vector of dependent variable
8  %              x : NxK matrix of independent variables
9  %              R : scalar replications
10 %------------------------------------------------
11 % OUTPUT: LL : scalar log likelihood
12 %              ll_i : log-likelihood contribution of
13 %                     observations
14 %------------------------------------------------
15
16 %----- (1) Seed and set up initial vectors ------
17 rng(1);
18 N               = size(y, 1);
19 Simulated_y     = NaN(N, R);
20
21 %----- (2) Create R simulated realizations ------
22 for count = 1:R
23     Simulated_y(:, count) =...
24                     SimulateBinaryLogit(x, Beta);
25 end
26 SimProb     = mean(Simulated_y, 2);
27
28 %----- (3) Calculate log-likelihood ------------
29 ll_i = y .* log(SimProb)+(1-y) .* log(1-SimProb);
30 LL     = -sum(ll_i);
31
32 return
```

We can now use `BinaryLogitSimulatedLL(Beta, y, x, R)` to estimate β:

```
>> R        = 1000;
>> opts2 = optimset('Algorithm', 'sqp',...
            'DiffMinChange', 1e-2);
>> [EstBetaMSL, LL, exitflag] =...
    fmincon(@(parameters) BinaryLogitSimulatedLL...
    (parameters, y, x, R), Beta_init, [], [], [],...
    [], lb, ub, [], opts2)

EstBetaMSL =

    0.5573
    0.6822

LL =

  601.5364

exitflag =

    2
```

There are several potential mysteries here worth exploring. First, on line 19, we create a matrix `Simulated_y`, which is full of elements that are 'not a number' (NaN). We will do this throughout the book whenever we are about to populate a matrix using a loop. We discuss the justification for doing so in Chapter 10. Second, notice that we fix the number of replications as $R = 1000$. You should play with this parameter: how does the performance of the estimator change as R varies? Third, notice that we increase the value of `DiffMinChange` from its default (zero). This is necessary because—as we will discuss in more detail soon— the simulated log-likelihood is locally flat. Finally, note that `BinaryLogitSimulatedLL` begins by seeding the random number generator—`rng(1)`. If we were to omit this, the simulated log-likelihood would generate slightly different values every time is called—even for the same data and the same parameters. This would be unnecessarily confusing—and likely to produce convergence problems when using `fmincon`.

4.2 Multinomial Logit

The previous discussion is a useful illustration of the principles of discrete binary choice. But, in many contexts, we want to model decision makers who face more than two alternatives. Generically, we can describe this as a model of 'multinomial choice'. For example, suppose that we now want to treat the decision maker as choosing between a set of different transport options. We now have an outcome variable with three values:

$$y_i = \begin{cases} 1 \text{ if the } i^{\text{th}} \text{ respondent drives to work;} \\ 2 \text{ if the } i^{\text{th}} \text{ respondent walks to work;} \\ 3 \text{ if the } i^{\text{th}} \text{ respondent cycles to work.} \end{cases} \quad (4.13)$$

There are several ways that we can model this problem. For example, we might be willing to assume that the choices are somehow *ordered*—that is, that the outcome varies monotonically in some latent variable. Alternatively, we may assume that the decision is *nested*—for example, we might treat commuters as deciding whether or not to drive and, if deciding not to drive, then deciding whether to walk or cycle.

In this chapter, we will eschew both of these assumptions. Instead, we will treat each outcome as generating utility with its own unobservable random component. For simplicity, we will maintain the 'Additive Random Utility Model' structure:

$$U_{ij}(x_i) = \alpha_{0j} + \alpha_{1j}x_i + \varepsilon_{ij}. \quad (4.14)$$

For choices $j \in \{1, 2, 3\}$, the individual obtains the following utilities:

$$U_{i1}(x_i) = \alpha_{01} + \alpha_{11}x_i + \varepsilon_{i1}; \quad (4.15)$$

$$U_{i2}(x_i) = \alpha_{01} + \alpha_{12}x_i + \varepsilon_{i2}; \quad (4.16)$$

$$U_{i3}(x_i) = \alpha_{03} + \alpha_{13}x_i + \varepsilon_{i3}. \quad (4.17)$$

Together, these three utilities determine the choice of an optimizing agent. Figure 4.1 illustrates preferences between the three options in two-dimensional space; in each box, the bold outcome represents the agent's choice.

We can express the conditional probability of the ith individual choosing, say, option 1 as:

$$\Pr(y_i = 1 \mid x_i) = \Pr[U_{i1}(x_i) > U_{i2}(x_i) \text{ and } U_{i1}(x_i) > U_{i3}(x_i) \mid x_i] \quad (4.18)$$

$$= \Pr[\alpha_{01} + \alpha_{11}x_i + \varepsilon_{i1} > \alpha_{02} + \alpha_{12}x_i + \varepsilon_{i2} \text{ and }$$

$$\alpha_{01} + \alpha_{11}x_i + \varepsilon_{i1} > \alpha_{03} + \alpha_{13}x_i + \varepsilon_{i3} \mid x_i]. \quad (4.19)$$

More generally, if the ith individual were to choose $y_i = j$ out of J choices, we could write:

$$\Pr(y_i = j \mid x_i) = \Pr\left[U_{ij}(x_i) > \max_{k \neq j}(U_{ik}(x_i)) \,\middle|\, x_i \right] \quad (4.20)$$

$$= \Pr\left[\alpha_{0j} + \alpha_{1j}x_i + \varepsilon_{ij} > \max_{k \neq j}(\alpha_{0k} + \alpha_{1k}x_i + \varepsilon_{ik}) \,\middle|\, x_i \right]. \quad (4.21)$$

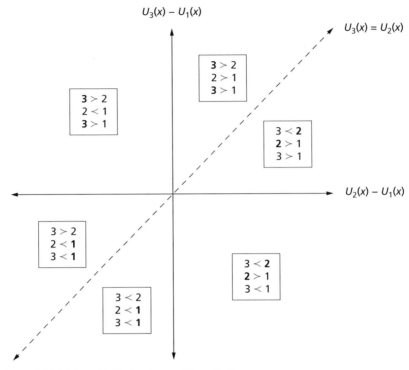

Figure 4.1 Multinomial Choice Among Three Options

To estimate using Equation 4.21, we will again assume that ε_{ij} is i.i.d. with a Type I Extreme Value distribution, and is independent of x_i. With this distributional assumption, we are able to find an expression for the conditional probability that the ith individual chooses outcome j from J choices.[3] Specifically, we can write:

$$\Pr(y_i = j \,|\, x_i) = \frac{\exp\left[\alpha_{0j} + \alpha_{1j}x_i\right]}{\sum_k \exp\left[\alpha_{0k} + \alpha_{1k}x_i\right]}. \tag{4.22}$$

It would be tempting to take this expression and immediately write the log-likelihood. For our three outcomes, for example, we could try to maximize the log-likelihood across six unknown parameters: $\alpha_{01}, \alpha_{02}, \alpha_{03}, \alpha_{11}, \alpha_{12}$, and α_{13}. However, this would be a mistake. We would be unable to find a unique set of values that maximize the function. That is, the model would be 'underidentified'. The reason for this is, of course, that utility only has meaning in *relative* terms. We therefore need to choose a 'base category', and estimate relative to the utility of that category. Let's choose option 1 as the base category, and define

---

[3] Train (2009, pages 74–5) shows a formal derivation of this result.

$\beta_{0j} \equiv \alpha_{0j} - \alpha_{01}$ and $\beta_{1j} \equiv \alpha_{1j} - \alpha_{11}$. (Of course, the choice of base category is arbitrary: our predicted probabilities would be identical if we were to choose a different base category.) With this assumption, we now multiply the numerator and the denominator of our conditional probability by $\exp(-\alpha_{01} - \alpha_{11}x_i)$ to obtain:

$$
\Pr(y_i = j \,|\, x_i) = \begin{cases} \dfrac{1}{1 + \sum_{k>1} \exp\left[\beta_{0k} + \beta_{1k}x_i\right]} & \text{for } j = 1; \\[4mm] \dfrac{\exp\left[\beta_{0j} + \beta_{1j}x_i\right]}{1 + \sum_{k>1} \exp\left[\beta_{0k} + \beta_{1k}x_i\right]} & \text{for } j > 1. \end{cases} \tag{4.23}
$$

Stacking the parameters into a single vector $\boldsymbol{\beta}$, we can now write the log-likelihood for the ith individual:

$$
\ell_i\left(\boldsymbol{\beta}; y_i \,|\, x_i\right) = \sum_{j=1}^{J} \mathbf{1}\left(y_i = j\right) \cdot \ln\left[\Pr(y_i = j \,|\, x_i)\right], \tag{4.24}
$$

where we use $\mathbf{1}(\cdot)$ to denote the indicator function. This log-likelihood function defines the famous 'Multinomial Logit' model. Note that, if $J = 2$, the Multinomial Logit collapses to the logit model that we considered earlier.

In Exercise (iii), you are asked to extend `SimulateBinaryLogit` to simulate the Multinomial Logit. Call this function `SimulateMNLogit(x, Betavec))`; like `SimulateBinaryLogit`, it should output both a vector of choices (`y`) and a matrix of simulated utilities (`utility`).[4] (Again, you can find our attempt on the companion website.) We can illustrate the simulation algorithm by graphing; it accepts as inputs the two matrices generated as outputs by `SimulateMNLogit`. Figure 4.2 shows the resulting cloud of simulated points, and the resulting simulated choices; this graph is drawn in $[U_2(x) - U_1(x), U_3(x) - U_1(x)]$ space, and is directly analogous to Figure 4.1.

GraphSimulatedData.m

```
1  function GraphSimulatedData(y, utility)
2  %-------------------------------------------------
3  % PURPOSE: Graph utility for a 3-good multinomial
4  %          choice
5  %-------------------------------------------------
6  % INPUTS:  utility: Nx3 matrix of utilities
7  %          y : Nx1 vector of choices
8  %-------------------------------------------------
9
```

---

[4] In `SimulateBinaryLogit`, `utility` is an $N \times 2$ matrix; in `SimulateMNLogit`, it should be an $N \times J$ matrix.

```
10  %----- (1) Clear axis and graph relative utilities
11  cla
12  scatter((utility(y==1,1) - utility(y==1,3)),...
13          (utility(y==1,2) - utility(y==1,3)))
14  hold on
15  scatter((utility(y==2,1) - utility(y==2,3)),...
16          (utility(y==2,2) - utility(y==2,3)))
17  hold on
18  scatter((utility(y==3,1) - utility(y==3,3)),...
19          (utility(y==3,2) - utility(y==3,3)))
20  hold on
21
22  %----- (2) Add boundary lines ------------------
23  plot([min((utility(:,1) - utility(:,3))), 0],...
24      [0, 0], 'LineWidth', 3, 'Color', 'k')
25  hold on
26  plot([0, max(utility(:, 1) - utility(:, 3))],...
27      [0, max(utility(:, 1) - utility(:, 3))],...
28      'LineWidth', 3, 'Color', 'k')
29  hold on
30  plot([0,0], [min(utility(:,1) - utility(:,3)),...
31      0], 'LineWidth', 3, 'Color', 'k')
32
33  return
```

Exercise (iv) asks you to extend BinaryLogitLL to generate a function returning $-1 \times \ell_i\left(\boldsymbol{\beta}; y_i \mid x_i\right)$ for the Multinomial Logit; you are then asked to use fmincon to maximize the log-likelihood and recover reasonable estimates of the original parameters. Exercise (v) asks you to do the same for the simulated log-likelihood—extending the function BinaryLogitSimulatedLL to produce a function called MNLogitSimulatedLL (and then to maximize).

4.3 Multinomial Probit

The Multinomial Logit is a very tractable model. As we have discussed, it provides an analytic expression for the log-likelihood; this function can therefore be evaluated and maximized easily. But this analytical tractability comes at a cost: the Multinomial Logit requires that the unobservable terms, ε_{ij}, have a Type I Extreme Value distribution, and that these terms are distributed independently of each other. This has serious implications for a structure of

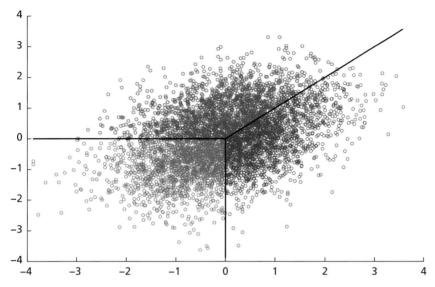

Figure 4.2 Simulating Choice

individual choice. Using Equation 4.22, we can write the ratio of the conditional probability that $y_i = A$ and that $y_i = B$:

$$\frac{\Pr\left(y_i = A \mid x_i\right)}{\Pr\left(y_i = B \mid x_i\right)} = \frac{\exp\left[\beta_{0A} + \beta_{1A}x_i\right]}{\exp\left[\beta_{0B} + \beta_{1B}x_i\right]} \tag{4.25}$$

$$= \exp\left[\beta_{0A} - \beta_{0B} + (\beta_{1A} - \beta_{1B}) \cdot x_i\right]. \tag{4.26}$$

Cameron and Trivedi (2005, p. 503) describe why this result is so concerning:

> As an extreme example, the conditional probability of commute by car given commute by car or red bus is assumed in [a Multinomial Logit] or [Conditional Logit] model to be independent of whether commuting by blue bus is an option. However, in practice we would expect introduction of a blue bus, which is the same as a red bus in every aspect except colour, to have little impact on car use and to halve use of the red bus, leading to an increase in the conditional probability of car use given commute by car or red bus.
>
> This weakness of MNL is known in the literature as the red bus—blue bus problem, or more formally as the assumption of independence of irrelevant alternatives.

This is clearly a serious limitation of the Multinomial Logit. Indeed, in his Nobel Prize Lecture in 2000, McFadden even went so far as to say (p. 339):

> The MNL model has proven to have wide empirical applicability, but as a theoretical model of choice behaviour its IIA property is unsatisfactorily restrictive.

In this section, we therefore consider a more flexible model: the Multinomial Probit. Let's rewrite our three-outcome model in terms of differences in utility, where we once again treat option 1 as the base category:

$$D_{i1}(x_i) \equiv U_{i1}(x_i) - U_{i1}(x_i) = 0; \tag{4.27}$$

$$D_{i2}(x_i) \equiv U_{i2}(x_i) - U_{i1}(x_i) = \beta_{02} + \beta_{12}x_i + \mu_{i2}; \tag{4.28}$$

$$D_{i3}(x_i) \equiv U_{i3}(x_i) - U_{i1}(x_i) = \beta_{03} + \beta_{13}x_i + \mu_{i3}. \tag{4.29}$$

Without loss of generality, we can now treat commuter i as choosing y_i as the maximum of $D_{i1}(x_i)$, $D_{i2}(x_i)$ and $D_{i3}(x_i)$. With this formulation of the model, we have reduced the dimensionality of our unobservable: we now have two error terms, rather than three.

In the Multinomial Logit, we made a distributional assumption over a J-dimensional vector: $(\varepsilon_{i1}, \varepsilon_{i2}, \varepsilon_{i3})$. In the Multinomial Probit, we will make a distributional assumption over a $(J-1)$-dimensional vector: (μ_{i2}, μ_{i3}).[5] Specifically, we assume that (μ_{i2}, μ_{i3}) has a Bivariate Normal distribution:

$$\begin{pmatrix} \mu_{i2} \\ \mu_{i3} \end{pmatrix} \sim \mathcal{N} \left[\begin{pmatrix} 0 \\ 0 \end{pmatrix}, \begin{pmatrix} 1 & \rho \\ \rho & 1 \end{pmatrix} \right]. \tag{4.30}$$

4.3.1 SIMULATING THE MODEL

The function SimulateMNProbit(x, Betavec, Omega) simulates the Multinomial Probit. Its operation is directly analogous to our simulation functions for the binary logit and for the Mulitnomial Logit.

SimulateMNProbit.m

```
1  function y = SimulateMNProbit(x, Betavec, Omega)
2
3  %------------------------------------------------
4  % PURPOSE: simulates the multinomial probit.
5  %------------------------------------------------
6  % INPUTS:   x : NxK vector of independent variable
7  %           Betavec : Kx1 parameter vector
8  %           Omega : covariance matrix
9  %------------------------------------------------
10 % OUTPUT:   choice : multinomial choice
11 %------------------------------------------------
12
```

[5] Train (2009, Chapter 5) shows how, with a multivariate normal error vector, this is equivalent to making an assumption in the J-dimensional space and then simplifying. But we will keep things simple and start with the $(J-1)$-dimensional space.

```
13   %----- (1) Setup initial vectors ----------------
14   N              = size(x, 1);
15   K              = size(x, 2);
16   J              = size(Betavec, 1)/K + 1;
17
18   %----- (2) Calculate relative utility ----------
19   Beta = reshape(Betavec, K, J - 1);
20   xb    = x * Beta;
21   diff = [zeros(N, 1), xb +...
22           mvnrnd(zeros(J - 1, 1), Omega, N)];
23
24   %----- (3) Based on relative utility, determine
25   %          choice from J -
26   [~, y] = max(diff, [], 2);
27
28   return
```

4.3.2 ESTIMATING BY SIMULATION

But wait! Why have we skipped straight to 'estimating by simulation'? Surely we have missed a step?! Exactly. We *have* missed a step—when we considered the binary logit and the Multinomial Logit, we (i) simulated our model, then (ii) estimated using Maximum Likelihood, and then (iii) estimated using maximum simulated likelihood. Step (iii) was redundant in both of these cases. But now we see why step (iii) matters—because in the case of Multinomial Probit, we cannot write the log-likelihood. That is, we can *simulate* our model, but we cannot use a closed-form expression to generate the log-likelihood. This is precisely the kind of problem that requires simulation.

The function MNProbitSimulatedLL(parameters, y, x, R) returns the simulated log-likelihood for the Multinomial Probit. In Exercise (vii), we ask you to use MNProbitSimulatedLL to recover the parameters for a simulated dataset.

MNProbitSimulatedLL.m

```
1   function LL = MNProbitSimulatedLL(param, y, x, R)
2
3   %-------------------------------------------------
4   % PURPOSE:  returns the simulated log-likelihood
5   % for the Multinomial Probit.
6   %-------------------------------------------------
```

```
7    % INPUTS:   x : Nx1 vector of independent variable
8    %           y : Nx1 vector of dependent variable
9    %           param : [Beta; Rho]
10   %           R : scalar replications
11   %-------------------------------------------------
12   % OUTPUT:    LL : log-likelihood
13   %-------------------------------------------------
14
15   %----- (1) Unpack initial parameters, Seed ------
16   BetaV           = param(1:end-1);
17   Rho             = param(end);
18   Omega           = [1, Rho; Rho, 1];
19   rng(1);
20
21   %----- (2) Create initial (prefilled) vectors ---
22   N               = size(y, 1);
23   J               = max(y);
24   Simulated_y     = NaN(N, R);
25   SimulatedProb   = NaN(N, J);
26
27   %----- (3) Simulate choices R times based on
28   %          current values---
29   for count = 1:R
30       Simulated_y(:, count) = SimulateMNProbit(x,...
31                            BetaV, Omega);
32   end
33
34   %----- (4) Create choice index matrix for each
35   %          item j in J ----
36   for j = 1:J
37    SimulatedProb(:, j) = mean(Simulated_y == j, 2);
38    MyIndex(:, j)       = (y == j);
39   end
40
41   %----- (5) Calculate maximum simulated
42   %          likelihood ----
43   ll_i     = sum(MyIndex .* log(SimulatedProb), 2);
44   LL       = -sum(ll_i);
45
46   return
```

4.3.3 A LOGIT-SMOOTHED AR SIMULATOR

So far, we have considered three functions for simulated log-likelihoods. In each case, we simulated a series of outcomes for each respondent i, $\{\tilde{y}_{i1}, \ldots, \tilde{y}_{iR}\}$, with the simulated $\Pr(y_i = j \mid x_i)$ calculated as:

$$\tilde{P}_j(x_i; \beta_0, \beta_1) \equiv \frac{1}{R} \cdot \sum_{r=1}^{R} \mathbf{1}\left(\tilde{y}_{ir} = j\right), \qquad (4.31)$$

where $\mathbf{1}(\cdot)$ is the indicator function. We can describe this as an 'accept-reject simulator'—for each replication, we simply decide whether or not the simulated outcome matches the outcome of interest.

This approach works—as our previous examples illustrated. However, it is far from perfect, for several reasons. First, the simulated log-likelihood is everywhere locally flat: it is a step function in the parameter space. The reason should be obvious: for a tiny change in the parameters, there is probability zero that any simulated outcome switches from 'accept' to 'reject' (or vice versa). For this reason, we needed to encourage MATLAB to take larger steps in its optimization (by changing the default option of DiffMinChange).

Second, for any finite number of replications R, there is no guarantee that, for a given individual, *any* simulated observation will match the observed choice. For example, in the binary logit case, it is possible that individual i chooses $y_i = 1$, but that all simulated outcomes for individual i are zero. Of course, this is more likely to occur when we are trying to evaluate the simulated log-likelihood far from the true parameter values, or when we are using a small number of replications. This is a fatal problem: we would conclude that $\tilde{P}_j(x_i; \beta_0, \beta_1) = 0$ for some observed outcome j, and will ask MATLAB to take the log of zero.

Third, the accept-reject simulator is throwing away information. When we simulate, we calculate values for the utility of each choice. However, the accept-reject simulator uses information only on which of those utility values is greatest—it discards useful information on whether an option was 'nearly chosen'. In this way, the accept-reject simulator is like a sports fan who wants to know whether his or her team won—but who does not care to hear the score.

These problems all arise because Equation 4.31 is discontinuous in the utility of option j—a small change in the simulated utilities can cause a discrete change in $\tilde{y}_{ir}$. We can therefore improve our simulator by smoothing over the simulated utilities. Following Train (2009, p. 121), denote the simulated utility for individual i for option j on replication r as U_{ij}^r. Then, instead of Equation 4.31, we can use:

$$\tilde{P}_j(x_i; \beta_0, \beta_1) \equiv \frac{1}{R} \cdot \sum_{r=1}^{R} \frac{\exp\left(U_{ij}^r / \phi\right)}{\sum_k \exp\left(U_{ik}^r / \phi\right)}. \qquad (4.32)$$

Equation 4.32 should look familiar: the term inside the summation is a Multinomial Logit transformation of the simulated utilities. For this reason, we can term this a 'logit-smoothed' simulator. The parameter ϕ is used to 'tune' the smoothness of the function. As $\phi \rightarrow \infty$, each simulated probability is smoothed to $1/J$; as $\phi \rightarrow 0$, the function approximates the (unsmoothed) accept-reject simulator. Train (p. 121) explains that there is 'little guidance on the appropriate level' for ϕ. As applied researchers, we need to experiment with this parameter in the context of our particular model.

The function MNProbitSimulatedLLSmoothed(parameters, y, x, R) implements the logit-smoothed Multinomial Probit. Compare the number of replications you need to those required in the unsmoothed case.

MNProbitSimulatedLLSmoothed.m

```matlab
1  function LL = ...
2  MNProbitSimulatedLLSmoothed(param, y, x, R)
3  %-------------------------------------------------
4  % PURPOSE: implements the logit-smoothed
5  %          multinomial probit
6  %-------------------------------------------------
7  % INPUTS: parameters : [Beta; Rho]
8  %           y : Nx1 vector of dependent variable
9  %           x : NxK matrix of independent variables
10 %           R : scalar simulate replications
11 %-------------------------------------------------
12 % OUTPUT:   LL : log-likelihood
13 %-------------------------------------------------
14
15 %----- (1) Unpack initial parameters, Seed ------
16 Betavec         = param(1:end-1);
17 Rho             = param(end);
18 Omega           = [1, Rho; Rho, 1];
19 rng(1);
20
21 %----- (2) Create initial (prefilled) vectors ---
22 N               = size(y, 1);
23 J               = max(y);
24 SimulatedS      = NaN(N, J, R);
25 SimulatedProb   = NaN(N, J);
26
27 %----- (3) Simulate (logit-smoothed) value
28 %                R times -----
29 for count = 1:R
```

```
30      SimulatedS(:, :, count) = ...
31        SimulateMNProbit_Smoothed(x, Betavec, Omega);
32    end
33    SimulatedProb    = mean(SimulatedS, 3);
34
35    %----- (4) Create choice index matrix for each
36    %              item j in J ----
37    for j = 1:J
38       MyIndex(:, j) = (y == j);
39    end
40
41    %----- (5) Calculate maximum simulated likelihood
42    ll_i       = sum(MyIndex .* log(SimulatedProb), 2);
43    LL         = -sum(ll_i);
44
45    return
```

Notice that MNProbitSimulatedLLSmoothed calls a function
SimulateMNProbitSmoothed. As the name implies, this is very similar
to SimulateMNProbit, but incorporates Equation 4.32 where we choose
$\phi = 0.01$, which seems to work well in this particular application.

SimulateMNProbitSmoothed.m

```
1    function S   =
2    SimulateMNProbitSmoothed(x, Betavec, Omega)
3    %-------------------------------------------------
4    % PURPOSE: implements the logit-smoothed
5    %              multinomial probit
6    %-------------------------------------------------
7    % INPUTS: x : NxK matrix of independent variables
8    %             Betavec : Kx1 parameter vector
9    %             Omega : covariance matrix
10   %-------------------------------------------------
11   % OUTPUT: S : smoothed choice
12   %-------------------------------------------------
13
14   %----- (1) Setup initial vectors and parameters -
15   phi       = 0.01;
16   N         = size(x, 1);
17   K         = size(x, 2);
18   J         = size(Betavec, 1)/K + 1;
19
```

```
20   %----- (2) Calculate relative utility, smooth ---
21   Beta     = reshape(Betavec, K, J - 1);
22   xb       = x * Beta;
23   diff     = [zeros(N, 1),xb +...
24               mvnrnd(zeros(J-1, 1), Omega, N)];
25   expdiff = exp(diff/phi);
26
27   %----- (3) Calculate smoothed choice for
28   %               realizations ---------
29   for count = 1:J
30       S(:, count) = expdiff(:, count) ./...
31                       sum(expdiff, 2);
32   end
33
34   return
```

4.4 **Review and Exercises**

Table 4.1 Chapter 4 Commands

Command	Brief description
cla	Clears current graph axes
end	Closes a for, if, or similar statement
	(Also represents the last location in a vector)
mvnrnd	Draws random numbers from the multivariate normal distribution
NaN	Pre-fills a matrix with 'empty' values
reshape	Moves elements column-wise from an initial matrix to a new matrix with specified dimensions
[~,...]	The tilde acts as a placeholder for an item which should not be returned

4.4.1 FURTHER READING

McFadden (2000) provides a fascinating overview of the history of discrete choice modelling, including a discussion of the famous Bay Area Rapid Transit analysis (initially reported in McFadden (1974)). Train (2009) provides an excellent discussion of discrete choice models; this book should be required reading for anyone with even a passing interest in the subject. Among many other topics, Train discusses in detail the Geweke-Hajivassiliou-Keane simulator, which is a further improvement on the smoothed accept-reject method. This simulator lies beyond the scope of our discussions here, but we encourage you to read Train's discussion of this method, or further details in Geweke et al. (1994) if you are interested in estimating Multinomial Probit models.

Maximum Simulated Likelihood is one very important class of simulation-based estimation, but certainly not the only one. Gourieroux and Monfort (1997) discuss simulation-based estimators in detail—including the Method of Simulated Moments and, more generally, Indirect Inference.

EXERCISES

(i) Go back and look again at the function SimulateBinaryLogit. How do the choice probabilites behave as the sample size N increases? How does the variance and mean of simulations change (numerically), as N grows? Are these results in line with what you would expect from the Central Limit Theorem and Law of Large Numbers?

(ii) In our estimations, $\hat{\beta}$ differs from β for two reasons: both our dataset and our simulation size are finite (ie $N < \infty, R < \infty$). Which is numerically more important? How does $\hat{\beta}$ respond as $N \to \infty$ for a fixed R? And as $R \to \infty$ for a fixed N?

(iii) Extend SimulateBinaryLogit to generate a function to simulate the Multinomial Logit. Call this function SimulateMNLogit(x, Betavec), where x is a scalar explanatory variable and betavec is a stacked vector of parameters. The function should output both a vector of choices (y) and a matrix of simulated utilities (utility).

(iv) Extend BinaryLogitLL to generate a function returning $-1 \times \ell_i \left(\beta; y_i \mid x_i \right)$ for the Multinomial Logit. Use fmincon to maximize the log-likelihood and recover reasonable estimates of the original parameters (that is, the parameters used as input Betavec to the function SimulateMNLogit).

(v) Extend BinaryLogitSimulatedLL to produce a function returning the simulated log-likelihood for the Multinomial Logit (called MNLogitSimulatedLL). Use fmincon to maximize the simulated log-likelihood and recover reasonable estimates of the original parameters (that is, the parameters used as input Betavec to the function SimulateMNLogit).

(vi) Both BinaryLogitLL and BinaryLogitSimulatedLL return the vector of ℓ_i as their second argument. Use this output to produce a scatterplot of ℓ_i under the analytical expression and under the simulated function. How does this scatterplot change as R changes?

(vii) Use MNProbitSimulatedLL and fmincon to recover the parameters for a dataset simulated using SimulateMNProbit.

(viii) Experiment with different values of phi in SimulateMNProbitSmoothed. Can you find a value of phi that performs better than $\phi = 0.01$?

5 Discrete Games

It is not a case of choosing those which, to the best of one's judgment, are really the prettiest, nor even those which average opinion genuinely thinks the prettiest. We have reached the third degree where we devote our intelligences to anticipating what average opinion expects the average opinion to be. And there are some, I believe, who practise the fourth, fifth and higher degrees.

Keynes*

Up to this point, we have considered observations on single agents—whether individual commuters, individual consumers, and so on. However, much microeconomic theory focuses upon strategic interactions between multiple players. It is useful to think about how we can model such interactions in MATLAB—and how we can estimate these models using data on the outcome of strategic interactions. In this chapter, we will study two-player binary games. We will first think about how to solve the game, and then about how we can use the solution for estimation. This will provide an intuitive introduction to the large literature on the structural estimation of strategic interactions.

5.1 A Simple Cournot Game

To get things started, let's discuss a familiar structure: a two-player Cournot game. Of course, this is not a discrete game (quantities are continuous)— but we will use this simple model to introduce the basic principles of solving games in MATLAB. Having established these principles, we turn to estimation in Section 5.2.

Suppose that we have two firms, each producing a single good, and that aggregate inverse demand is given as follows (also illustrated in Figure 5.1):

$$p(q) = \begin{cases} a - b \cdot q & \text{for } q < a \cdot b^{-1}; \\ 0 & \text{for } q \geq a \cdot b^{-1}. \end{cases} \tag{5.1}$$

We assume that each firm i chooses a quantity to produce, $q_i \geq 0$. Actions are taken simultaneously. For a combination of quantities (q_i, q_j), the payoff (profit) to firm i is:

---

* Keynes, J. M., *The General Theory of Employment Interest and Money*, Palgrave Macmillan (1936).

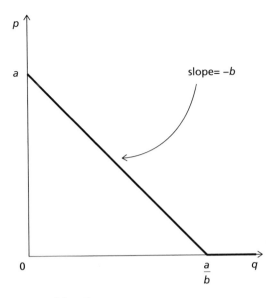

Figure 5.1 Inverse Demand Function

$$\pi_i(q_i, q_j) = \left[a - b \cdot (q_i + q_j) \right] \cdot q_i - c \cdot q_i \tag{5.2}$$

$$= (a - c - b \cdot q_j) \cdot q_i - b \cdot q_i^2. \tag{5.3}$$

5.1.1 FINDING THE BEST-RESPONSE FUNCTION

We want to find the Nash Equilibrium. To do this, we need the best-response functions. Trivially, the best-response for firm i is solved as:

$$\left. \frac{\partial \pi_i(q_i, q_j)}{\partial q_i} \right|_{q_i = q_i^*} = a - c - b \cdot q_j - 2b \cdot q_i = 0 \tag{5.4}$$

$$\therefore q_i^*(q_j) = \frac{a - c - b \cdot q_j}{2b} = \frac{a - c}{2b} - \frac{q_j}{2}. \tag{5.5}$$

Often, we are interested in solving games for which there is no analytical solution—indeed, we will consider one such example later in this chapter. So that you are equipped with the tools necessary to solve such games, let's assume that we cannot find the analytical solution of Equation 5.5. How might we solve this game numerically in MATLAB? Let's build the problem from its core parts. First, we need functions for the aggregate inverse demand, $p(q; a, b)$, and for the profit function of firm i, $\pi_i (q_i, q_j; c_i, a, b)$. Exercises (i) and (ii) invite you to code these functions, which we will call simply Demand and Profit respectively. These functions are also available on our accompanying website.

We need to find firm i's best-response, q_i^*, numerically. Remember the key idea to the best-response concept: we treat firm i as optimizing subject to firm j's decision, q_j. That is, we solve:

$$q_i^*(q_j) = \arg\max_{q_i \geq 0} \pi_i(q_i, q_j). \tag{5.6}$$

It is straightforward to implement this in MATLAB using fmincon. We nest this optimization in a function called BestResponse:

BestResponse.m

```
1   function qi        = BestResponse(qj, ci, a, b)
2   %--------------------------------------------------------
3   % PURPOSE:  returns the best response for firm i to
4   %              output q_j.
5   %--------------------------------------------------------
6   % INPUTS:    qj : quantity produced by firm j
7   %            ci : MC for i
8   %            a : intercept
9   %            b : slope
10  %--------------------------------------------------------
11  % OUTPUT:    qi : best reponse for firm i
12  %--------------------------------------------------------
13
14  options = optimset('Algorithm', 'sqp',...
15                 'Display', 'off');
16  qi      = fmincon(@(q) -Profit(q, qj, ci, a, b),...
17              0, [], [], [], [], 0, [], [], options);
18
19  return
```

The function BestResponse returns a single value for q_i^*, for a single given value for q_j. We can visualize this as a function by looping over a vector of possible values for q_j. The following code shows how this is done, and Figure 5.2 illustrates the result.

GraphBestResponse.m

```
1   %----(1) Initialize the parameters, generate grid
2   %            for qj -----
3   a    = 20;
4   c    = 2;
```

```
5   b    = 1;
6   qj   = [0:20]';
7
8   %--- (2) Solve the best response for qi...  ------
9   qi_star      = NaN(size(qj));
10
11  for count = 1:size(qj, 1)
12      qi_star(count, 1) = BestResponse(qj(count,...
13                          1), c, a, b);
14  end
15
16  line(qj, qi_star, 'LineWidth', 2, 'Color',...
17      [1, 0, 0])
18  ylabel('q_i', 'FontSize',12)
19  xlabel('q_j', 'FontSize', 12)
```

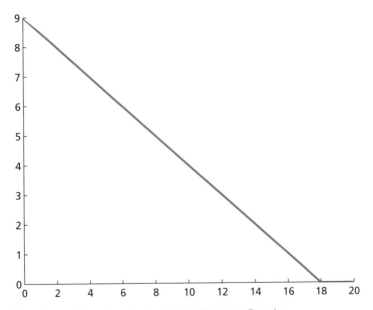

Figure 5.2 A Numeric Solution for Firm i's Best Response Function

So far, so good. But, of course, what we really want is to overlay the *two* best-response functions: $q_i^*(q_j)$ and $q_j^*(q_i)$. Having coded the function BestResponse in such general terms, this is easy: we simply repeat the method we used to graph $q_i^*(q_j)$, switching i and j ....

GraphNashEquilibrium.m

```
1  %---- (3) Generate a grid for qi...--------------
2  qi   = [0:20]';
3
4  %--- (4) Solve the best response for qj...-------
5  qi_star      = NaN(size(qj));
6
7  for count = 1:size(qj, 1)
8      qj_star(count, 1) = BestResponse(qi(count,...
9                          1), c, a, b);
10 end
```

We can now overlay this function on our earlier graph (ensuring that, as in the earlier case, we are graphing q_j on the x axis and q_i on the y axis). Figure 5.3 illustrates.

```
1  hold on
2  line(qj_star, qi, 'LineWidth', 2, 'Color',...
3      [0, 0, 1])
```

From these two functions, it is clear that the Nash Equilibrium is $(6, 6)$—where the two best response functions intersect. You should check that you obtain the same solution analytically by using Equation 5.5.

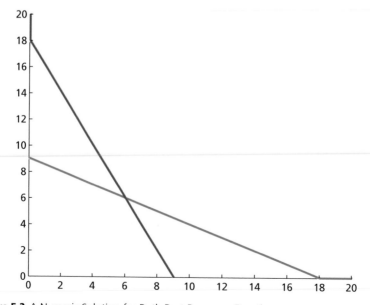

Figure 5.3 A Numeric Solution for Both Best Response Functions

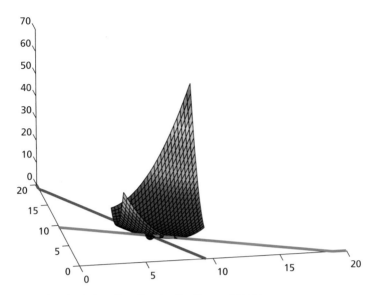

Figure 5.4 A Quadratic Loss Function for the Cournot Game

However, we are not done yet—if we want to use our model for estimation, we need a way for MATLAB to calculate this equilibrium automatically, rather than have us eyeball graphs of the output. We can do this by defining a simple loss function. Suppose that our model generates best-response functions $(q_i^*(j), q_j^*(i))$. Then for some pair (q_i, q_j), we can define the loss as:

$$L(q_i, q_j) = \left(q_i - q_i^*(q_j)\right)^2 + \left(q_j - q_j^*(q_i)\right)^2. \tag{5.7}$$

$L(q_i, q_j)$ is uniquely minimized (at zero) when (q_i, q_j) is a Nash Equilibrium.

The function CournotLoss returns the value of this loss given the best response functions in MATLAB. Figure 5.4 shows the loss function for the region around the Nash. You can reproduce this graph using the methods that we discussed in Chapter 1, including the meshgrid command. The Nash can be found numerically simply by minimising CournotLoss subject to the constraint that quantities are non-negative. This is done by the function SolveCournotNash.

CournotLoss.m

```
1  function Loss    = CournotLoss(q, a, b, ci, cj)
2  %-------------------------------------------------
3  % PURPOSE: returns the quadratic loss for a
4  % simple Cournot model.
5  %-------------------------------------------------
```

```
 6  % INPUTS:     q : [q(1); q(2)]
 7  %             ci : MC for i
 8  %             cj : MC for j
 9  %             a : intercept
10  %             b : slope
11  %-----------------------------------------------
12  % OUTPUT:   Loss : deviation from best response
13  %-----------------------------------------------
14
15  qi_star = BestResponse(q(2), ci, a, b);
16  qj_star = BestResponse(q(1), cj, a, b);
17
18  Loss    = (q(1) - qi_star)^2 + (q(2) - qj_star)^2;
19
20  return
```

SolveCournotNash.m

```
 1  function Nash = SolveCournotNash(a, b, ci, cj)
 2  %-----------------------------------------------
 3  % PURPOSE: solves the Nash  for a simple Cournot
 4  %          model.
 5  %-----------------------------------------------
 6  % INPUTS:   ci : MC for i
 7  %           cj : MC for j
 8  %           a : intercept
 9  %           b : slope
10  %-----------------------------------------------
11  % OUTPUT:   Nash : quantites at equilibrium
12  %-----------------------------------------------
13
14  opts =   optimset('Algorithm', 'sqp',...
15           'Display', 'off');
16  Nash =   fmincon(@(q) CournotLoss(q, a, b, ci,...
17           cj), [0, 0], [], [], [], [], [0, 0],...
18           [], [], opts);
19
20  return
```

Let's use `SolveCournotNash` to find the Nash Equilibrium, and add it (as a large black dot) to our earlier graph:

```
>> Nash = SolveCournotNash(a, b, cj, ci)

Loss =

    1.2986e-14

exitflag =

    1

Nash =

    6.0000    6.0000

>> hold on
>> scatter(Nash(1), Nash(2), 100, [0, 0, 0],...
   'filled')
```

Notice that each of our functions allows for asymmetric cost, c_i and c_j. You should experiment: how does cost asymmetry change the Nash? In Exercises (iii) and (iv), we ask you to go further—to adapt the code to an N-firm Cournot game, and then a two-firm Stackelburg game.

5.2 A Discrete Bayesian Game

We just showed how we can use standard optimization methods to find a Nash Equilibrium. We did this by minimizing over a joint loss function, where loss is defined as the sum of the profit deviation between each firm's action and that firm's best response. We will now apply these principles for estimation.

When we estimate models of individual decision making, we have multiple observations, each showing the result of an *individual* decision. (For example, when we model the decision of a commuter whether or not to drive to work, we use data on multiple commuters and their decisions.) When we model games, we need to observe a *game* being played multiple times.

We will now consider a stylized Bayesian game. Specifically, we consider a simple two-firm coordination game. Let's assume that we have data on markets $m \in \{1, \ldots, M\}$. In each market, we observe competition between the same two large firms (which we call 'Firm 1' and 'Firm 2'). In each market, each firm needs to decide whether to launch a local radio advertising campaign ($a_i = 1$) or not ($a_i = 0$).[1]

---

[1] This empirical motivation is inspired in part by Sweeting (2009), who uses a structural approach to model US radio stations' strategic decisions on the timing of advertising.

We assume that profit accrues through two sources: (i) firm i receives a signal, x_i, which records the firm's net profit from choosing the radio campaign, and (ii) firm i receives an additional payoff δ_i if firm i and firm j make the same choice (i.e. if $a_i = a_j$). In this context, we should think of δ_i and δ_j as measuring a relative payoff to a firm for differentiating itself from its competitor. Specifically, we will assume that (ceteris paribus) a firm weakly prefers its competitor *not* to be using the same advertising strategy: $\delta_i, \delta_j \leq 0$. We assume that the firms act simultaneously.

To fix ideas, let's focus on solving the game in a particular market. In normal form, we can express the game as follows:

		FIRM 1 (a_1)	
		0	1
FIRM 2 (a_2)	0	δ_2, δ_1	$0, x_1$
	1	$x_2, 0$	$x_2 + \delta_2, x_1 + \delta_1$

Firm i's payoff is then:

$$U_i(a_i; a_j, x_i) = \begin{cases} x_i + \delta_i & \text{if } a_i = 1, a_j = 1; \\ x_i & \text{if } a_i = 1, a_j = 0; \\ 0 & \text{if } a_i = 0, a_j = 1; \\ \delta_i & \text{if } a_i = 0, a_j = 0. \end{cases} \tag{5.8}$$

We assume that the researcher observes (a_1, a_2), but that x_1 and x_2 are only observable to Firm 1 and Firm 2 respectively. That is, each firm forms its own assessment of the relative profit from using radio advertising, and this is shared neither with its competitor firm nor with the researcher. We therefore have a game of incomplete information: only Firm 1 knows x_1, and only Firm 2 knows x_2. In this model, (x_1, x_2) acts like an error term: it generates variation in the outcome that is not explained by the measured covariates. As in the previous chapter, we need a distributional assumption on this unobservable. For simplicity, we will use the Bivariate Normal:

$$\begin{pmatrix} x_1 \\ x_2 \end{pmatrix} \sim \mathcal{N}\left(\begin{pmatrix} \mu_1 \\ \mu_2 \end{pmatrix}, \begin{pmatrix} 1 & \rho \\ \rho & 1 \end{pmatrix} \right). \tag{5.9}$$

We will use the notation $\Phi_2(\mu_1, \mu_2, \rho)$ for the *cdf* of a Bivariate Normal with this structure.

Given this assumption, we can solve the model. For firm i, the expected profit of $a_i = 1$—conditional on the firm observing x_i—is:

$$U_i(1; 1, x_i^*) \cdot \Pr\left(a_j = 1 \mid x_i\right) + U_i(1; 0, x_i^*) \cdot \left(1 - \Pr\left(a_j = 1 \mid x_i\right)\right). \tag{5.10}$$

The expected utility of $a_i = 0$ (again, conditional on observing x_i) is:

$$U_i(0; 1, x_i^*) \cdot \Pr\left(a_j = 1 \mid x_i\right) + U_i(0; 0, x_i^*) \cdot \left(1 - \Pr\left(a_j = 1 \mid x_i\right)\right), \tag{5.11}$$

where x_i^* as the value of x_i that equalizes Equations 5.10 and 5.11.

It is easy to see that the relative utility from choosing $a_i = 1$ is increasing in x_i. We therefore limit attention to a simple cut-off strategy:[2]

$$a_i = \begin{cases} 0 \text{ if } x_i < x_i^*; \\ 1 \text{ if } x_i \geq x_i^*. \end{cases} \tag{5.12}$$

So how, then, should firm i choose its cutoff value x_i^*? We know that, if $x_i = x_i^*$, firm i is indifferent between $a_i = 0$ and $a_i = 1$. We can therefore say:

$$U_i(1; 1, x_i^*) \cdot \Pr\left(a_j = 1 \mid x_i = x_i^*\right) + U_i(1; 0, x_i^*) \cdot \left(1 - \Pr\left(a_j = 1 \mid x_i = x_i^*\right)\right)$$
$$= U_i(0; 1, x_i^*) \cdot \Pr\left(a_j = 1 \mid x_i = x_i^*\right) + U_i(0; 0, x_i^*) \cdot \left(1 - \Pr\left(a_j = 1 \mid x_i = x_i^*\right)\right) \tag{5.13}$$

$$\therefore x_i^* = \delta_i \cdot \left(1 - 2\Pr\left(a_j = 1 \mid x_i = x_i^*\right)\right) \tag{5.14}$$

$$= \delta_i \cdot \left(2\Pr\left(a_j = 0 \mid x_i = x_i^*\right) - 1\right) \tag{5.15}$$

Equation 5.15 shows that the signal x_i plays two roles: (i) it contributes directly to firm i's utility, and (ii) it provides information to firm i on the likely action of firm j.

From the assumption of Bivariate Normality, we know:

$$x_j \mid x_i \sim \mathcal{N}\left(\mu_j + \rho \cdot (x_i - \mu_i), 1 - \rho^2\right) \tag{5.16}$$

$$\therefore \left. \frac{x_j - \mu_j - \rho \cdot (x_i - \mu_i)}{\sqrt{1 - \rho^2}} \right| x_i \sim \mathcal{N}(0, 1). \tag{5.17}$$

Define x_j^* as the cutoff point for Firm j. From this, Firm i can reason as follows:

$$\Pr(a_j = 1 \mid x_i) = \Pr(x_j \geq x_j^* \mid x_i) \tag{5.18}$$

$$= \Pr\left(\left. \frac{x_j - \mu_j - \rho \cdot (x_i - \mu_i)}{\sqrt{1 - \rho^2}} \geq \frac{x_j^* - \mu_j - \rho \cdot (x_i - \mu_i)}{\sqrt{1 - \rho^2}} \right| x_i \right) \tag{5.19}$$

$$= 1 - \Phi\left(\frac{x_j^* - \mu_j - \rho \cdot (x_i - \mu_i)}{\sqrt{1 - \rho^2}}\right) \tag{5.20}$$

$$\therefore \Pr(a_j = 0 \mid x_i) = \Phi\left(\frac{x_j^* - \mu_j - \rho \cdot (x_i - \mu_i)}{\sqrt{1 - \rho^2}}\right). \tag{5.21}$$

---

[2] Limiting attention to cutoff strategies in this way is common in the literature on Bayesian games—see, for example, Morris and Shin (2003). See Xu (2014) for further discussion of the use of cutoff strategies in a two-player binary choice model with correlated unobservables.

Therefore, the cut-off value for firm i's best-response function is defined by:

$$x_i^* = \delta_i \cdot \left[2 \cdot \Phi \left(\frac{x_j^* - \mu_j - \rho \cdot (x_i^* - \mu_i)}{\sqrt{1 - \rho^2}} \right) - 1 \right]. \qquad (5.22)$$

Equation 5.22 characterizes the equilibrium. Specifically, it describes a Bayesian Nash Equilibrium. It is sufficient for the equilibrium to be unique that:[3]

$$\delta_i, \delta_j > -\sqrt{\frac{\pi}{2}} \cdot \sqrt{\frac{1 - \rho}{1 + \rho}}. \qquad (5.23)$$

You should check that you understand that intuition behind these best-response functions. For example, what should firm i do if it has no preference for coordination (that is, $\delta_i = 0$)? What happens if the signals (x_i, x_j) are independent, conditional on (μ_i, μ_j) (that is, if $\rho = 0$)? Given a set of parameters, what are the maximum and minimum possible values for x_i^*? Try varying δ: can you find multiple equilibria by violating Equation 5.23? (Note: Remember that Equation 5.23 is *sufficient* for a unique equilibrium, but not *necessary*— you should be able to find cases in which Equation 5.23 is violated but where the equilibrium is still unique.[4])

Figure 5.5 shows a numerical solution to this binary game, for the parameter values $\mu_i = 1$, $\mu_j = -1$, $\rho = 0.75$ and $\delta_i = \delta_j = -0.5$. You now have the tools to produce this figure; this is directly analogous to the process that we just followed for the Cournot game—albeit with a more complicated equation defining the best responses. Exercises (v) to (viii) invite you to walk through each step of this numerical process.

5.2.1 SIMULATING THE GAME

How can we take this model to data? We will use a simple parameterization. We assume that we observe a series of independent markets, indexed $m \in \{1, \ldots, M\}$. In each market, we will be able to identify one firm as 'Firm 1' and the other as 'Firm 2'. We assume that, for each firm in each market, we observe a single binary covariate, z_{im}, and that this covariate is common knowledge between firms. We will also assume that we know the size of each market (measured in millions of consumers, rounded to the nearest 0.1); we denote this n_m. Our data, therefore, looks something like this...

---

[3] This is adapted from a condition in Morris and Shin (2003). It ensures that, after their intersection, the best-response functions for Firm 1 and Firm 2 never turn back towards each other. Gole and Quinn (2014) derive the equivalent condition for the more complicated three-player case, and discuss the relationship to this two-player setup.

[4] That said, Equation 5.23 provides the *weakest* sufficient condition—in the sense that we *can* find multiple equilibria if the condition is violated even by an infinitesimal margin.

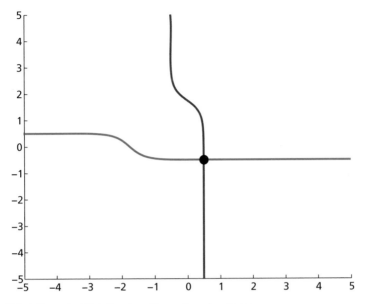

Figure 5.5 A Numerical Solution to a Binary Bayesian Game

		FIRM 1		FIRM 2	
m	n_m	z_{1m}	a_{1m}	z_{2m}	a_{2m}
1	0.3	0	0	0	1
2	0.1	1	0	0	0
$\vdots$	$\vdots$	$\vdots$	$\vdots$	$\vdots$	$\vdots$
M	0.9	1	1	0	0

In this example, imagine that variation in the binary covariate z_{im} causes variation in the preference for coordination:

$$\delta_{1m} = \beta_1 \cdot z_{1m}; \tag{5.24}$$

$$\delta_{2m} = \beta_2 \cdot z_{2m}. \tag{5.25}$$

Under this structure, we therefore observe some markets in which $(\delta_{1m}, \delta_{2m}) = (0, 0)$ where $(z_{1m}, z_{2m}) = (0, 0)$ This is important for identification.[5] Further, we will assume that the mean payoff to advertising for an individual firm is a linear function of market population:

$$\mu_{1m} = \gamma_1 \cdot n_m; \tag{5.26}$$

$$\mu_{2m} = \gamma_2 \cdot n_m. \tag{5.27}$$

---

[5] We say almost nothing about identification in this book, and we do not intend to start now. The further readings discussed at the end of this chapter involve much discussion of identification of discrete games.

ρ here is common across markets. Therefore, we have five parameters to estimate, which we stack in a vector: $\boldsymbol{\theta} = (\gamma_1, \gamma_2, \beta_1, \beta_2, \rho)'$.

First, let's simulate the model to generate a dataset to use for estimation. The function SimulateBayesianNash simulates values for n_m, z_{1m}, z_{2m}, a_{1m} and a_{2m}. (The function calls SolveBayesianNash, which you are asked to code yourself in Exercise (vii). Alternatively, you can download it from our accompanying website.)

SimulateBayesianNash.m

```
function [a,z,nm]...=
SimulateBayesianNash(M,g_1,g_2,b_1,b_2,rho);
%------------------------------------------------
% PURPOSE: simulates data from the Bayesian Nash
%          model.
%------------------------------------------------
% INPUTS:   M : total markets
%           g_1 : gamma 1 (parameter)
%           g_2 : gamma 2 (parameter)
%           b_1 : beta 1 (parameter)
%           b_2 : beta 2 (parameter)
%           rho : signal correlation (parameter)
%------------------------------------------------
% OUTPUT:   a : advertising strategy choice
%           z : binary covariate
%           nm : market size
%------------------------------------------------
rng(1)

%--- First, simulate the covariates (z and p).---
z       = randi([0, 1], M, 2);
nm      = randi([0, 10], M, 1)/ 10;

%---- Second, simulate by solving the game for
%        each market...----
cutoffs = NaN(M, 2);
a       = NaN(M, 2);

for count = 1:M
    count
    cutoffs(count, :) =...
    SolveBayesianNash(g_1 * nm(count),...
                      g_2 * nm(count),...
                      b_1 * z(count, 1),...
```

```
35                            b_2 * z(count, 2),...
36                            rho);
37        a(count, :)    = mvnrnd([g_1 * nm(count),...
38                            g_2 * nm(count)],...
39                            [1, rho; rho, 1], 1) > ...
40                            cutoffs(count, :);
41   end
42
43   return
```

Let's use this function to simulate a dataset of $M = 500$ markets . . .

```
>> gamma_1          = 1;
>> gamma_2          = 0.5;
>> beta_1           = -0.4;
>> beta_2           = -0.7;
>> rho              = 0.5;
>> [a, z, nm]       = SimulateBayesianNash(500,...
                      gamma_1, gamma_2,...
                      beta_1, beta_2, rho);
```

5.2.2 ESTIMATION

Denote (x_{1m}^*, x_{2m}^*) as the solution to the game in market m—that is, the pair of cutoffs that defines the Bayesian Nash Equilibrium. We have that:

$$\Pr(a_1 = 0, a_2 = 0) = \Phi_2\left(x_1^* - \mu_1, x_2^* - \mu_2, \ \rho\right); \tag{5.28}$$

$$\Pr(a_1 = 1, a_2 = 0) = \Phi_2\left(\mu_1 - x_1^*, x_2^* - \mu_2, -\rho\right); \tag{5.29}$$

$$\Pr(a_1 = 0, a_2 = 1) = \Phi_2\left(x_1^* - \mu_1, \mu_2 - x_2^*, -\rho\right); \tag{5.30}$$

$$\Pr(a_1 = 1, a_2 = 1) = \Phi_2\left(\mu_1 - x_1^*, \mu_2 - x_2^*, \ \rho\right). \tag{5.31}$$

Therefore, for a market m, the log-likelihood is:

$$\ell\left(\boldsymbol{\theta}; a_{1m}, a_{2m} \mid z_{1m}, z_{2m}, n_m\right)$$
$$= \ln \Phi_2\left\{\left[(2a_{1m} - 1) \cdot \left(\gamma_1 z_{1m} - x_{1m}^*\right), (2a_{2m} - 1) \cdot \left(\gamma_2 z_{2m} - x_{2m}^*\right),\right.\right.$$
$$\left.\left.(2a_{1m} - 1) \cdot (2a_{2m} - 1) \cdot \rho\right]\right\}. \tag{5.32}$$

The function BayesianNashLL implements this. Notice that the function loops over the elements of a matrix called 'UniqueData'. Why? What's the advantage of this approach?

BayesianNashLL.m

```
1   function LL = BayesianNashLL(param, a, z, nm)
2
3   %-----------------------------------------------
4   % PURPOSE: log-likelihood of Bayesian Nash
5   %          equilibrium
6   %-----------------------------------------------
7   % INPUTS:   a : advertising strategy choice
8   %           z : binary covariate
9   %           nm : market size
10  %           param : [gamma; beta]
11  %-----------------------------------------------
12  % OUTPUT:   LL : log-likelihood
13  %-----------------------------------------------
14
15  g_1        = param(1);
16  g_2        = param(2);
17  b_1        = param(3);
18  b_2        = param(4);
19  rho        = param(5);
20
21  N          = size(a, 1);
22
23  [UniqueData, m, n]   = unique([z, nm], 'rows');
24  cutoffs_small        = NaN(size(UniqueData, 1),...
25                           2);
26
27  for c = 1:size(UniqueData, 1)
28      cutoffs_small(c, :) = SolveBayesianNash(...
29                          g_1*UniqueData(c, 3),...
30                          g_2*UniqueData(c, 3),...
31                          b_1*UniqueData(c, 1),...
32                          b_2*UniqueData(c, 2),...
33                          rho);
34  end
35
36  cutoffs              = cutoffs_small(n, :);
37  ll                   = NaN(N, 1);
38
39  for c = 1:N
40      q1    = 2 * a(c, 1) - 1;
41      q2    = 2 * a(c, 2) - 1;
```

```
42
43      ll(c) = log(bvnl(q1 * (g_1 * nm(c,1) -...
44          cutoffs(c,1)), q2 * (g_2 * nm(c,2) -...
45          cutoffs(c,2)), q1 * q2 * rho));
46
47  end
48
49  LL   = -sum(ll);
50
51  return
```

We can recover estimates of θ by maximizing this log-likelihood. However, before we do, we have one more problem to solve. Remember that, if δ_{1m} or δ_{2m} are sufficiently large negative numbers, our game will have multiple solutions. This poses a problem for our estimation—because we have not specified any rule by which firms choose between multiple equilibria. There are certainly ways that we could deal with this—for example, we could impose an equilibrium selection rule, or we could even assume that firms somehow jointly mix across multiple equilibria with some predetermined probability (see, for example, Sweeting (2009) and de Paula and Tang (2012)). But, to keep things simple for now, we would like to rule out the possibility of multiple equilibria.

Recall that Equation 5.23 provides a sufficient condition for equilibrium uniqueness. We can impose this condition as a non-linear constraint on our maximum likelihood algorithm. (Of course, note that Equation 5.23 is a condition on δ_{1m} and δ_{2m} in a single game. In our empirical implementation, we have $\delta_{1m} = \beta_1 \cdot z_{1m}$ and $\delta_{2m} = \beta_2 \cdot z_{2m}$, with z_{1m} and z_{2m} each being dummy variables. Therefore, we can impose the constraint in Equation 5.23 as if it applies to β_1 and β_2, rather than to δ_1 and δ_2.)

We do this in the function `UniqueEquilibriumConstraint`:

UniqueEquilibriumConstraint.m

```
1   function [c, ceq] =
2   UniqueEquilibriumConstraint(param)
3   %---------------------------------------------
4   % PURPOSE: non-linear constraint for uniqueness
5   %---------------------------------------------
6   % INPUTS:    param : [gamma; beta; rho]
7   %---------------------------------------------
8   % OUTPUT:    c : non-linear inequality constraint
9   %              ceq : non-linear equality constraint
10  %---------------------------------------------
```

```
11
12 beta_1   = param(3);
13 beta_2   = param(4);
14 rho      = param(5);
15
16 c    = [-sqrt(pi/2) * sqrt((1 - rho) /...
17          (1 + rho)) - beta_1; -sqrt(pi/2) *...
18          sqrt((1 - rho) / (1 + rho)) - beta_2];
19 ceq  = [];
20
21 return
```

UniqueEquilibriumConstraint returns two outputs: c and ceq. This is necessary for fmincon as it looks for two types of non-linear constraint:

$$c \leq 0; \tag{5.33}$$

$$ceq = 0. \tag{5.34}$$

In this case, we have a vector of two values that must be negative (a restriction on δ_1 and on δ_2). These restrictions are captured by c. We have no non-linear equality constraint, so we set ceq to an empty matrix.

We can now maximize, imposing the unique-equilibrium constraint:

```
>> parameters_init  = [gamma_1, gamma_2, beta_1,...
                       beta_2, rho]';
>> lb    = [-2, -2, -2, -2, 0];
>> ub    = [2, 2, 0, 0, 0.99];
>> opts  = optimset('Algorithm', 'sqp',...
              'Display', 'iter', 'DiffMinChange',...
              1e-4);

>> [param, LL, exitflag] =...
   fmincon(@(par) BayesianNashLL(par, a, z, nm),...
         parameters_init, [], [], [], [], lb, ub,...
         @(par) UniqueEquilibriumConstraint(par),...
         opts);

>> [parameters_init, param]

ans =

       1.0000     1.0041
       0.5000     0.5405
      -0.4000    -0.3808
```

```
-0.7000    -0.6416
 0.5000     0.5711
```

The algorithm performs well: we recover good estimates of all parameters. For completeness, we should check that, as required, our estimated parameters obey the single-equilibrium condition:

```
>> UniqueEquilibriumConstraint(param)

ans =

    -0.2740
    -0.0132
```

As required, all elements of the vector c are negative. Of course, the only way for either element not to be negative would be if we had made a coding mistake—or if fmincon had not successfully converged. This result tells us nothing about whether there actually are multiple equilibria in the dataset we study—it merely tells us that we have estimated while successfully imposing the constraint in Equation 5.23. In this case, we are using simulated data, so can check the unique equilibrium condition with the true parameters:

```
>> UniqueEquilibriumConstraint(parameters_init)

ans =

    -0.3236
    -0.0236
```

5.3 **Review and Exercises**

Table 5.1 Chapter 5 Commands

Command	Brief Description
unique	Selects only non-repeating values
nonlcon	Non-linear contraints to be used in optimizers. Should be passed as a function, and list both equality and inequalities.

5.3.1 FURTHER READING

The basic structure considered in this chapter is broadly similar to that of an entry game. The econometrics of entry games were famously studied in seminal work by Bresnahan and Reiss (1990, 1991a,b) and Berry (1992) (see also Heckman (1978), Tamer (2003) and Ciliberto and Tamer (2009)). Such models typically use a 'complete information' structure—as Bresnahan and Reiss (1991a, p. 59) explain, 'Players of the game observe [the realisation of a random variable affecting payoffs] and have complete information about other players' actions. The econometrician, however, does not observe players' payoffs and therefore treats them as random variables.'

A related literature considers cases where—as in this chapter—players have incomplete information about each other's payoffs. This literature tends to assume that players' idiosyncratic signals are independent of each other (conditional on variables observable to the researcher): see, in particular, Sweeting (2009), Bajari et al. (2010), and de Paula and Tang (2012). More recently, Grieco (2014) uses a flexible parametric structure to combine 'public shocks' (observable to players but not the researcher) and 'private shocks' (each observable only to its respective player); this can nest a complete information game as a limiting case. Gole and Quinn (2014) follow this literature in estimating three-player binary games with correlated signals; the model presented in this chapter is a two-player simplification of their approach. (See also Xu (2014) and Wan and Xu (2014), who consider a related model with correlated unobservables and continuous covariates.)

EXERCISES

(i) Write a function, `Demand`, to return the aggregate inverse demand for the simple Cournot model in Section 5.1, $p(q; a, b)$.

(ii) Write a function, `Profit`, to return the profit of firm i in the simple Cournot model in Section 5.1, $\pi_i(q_i, q_j; c_i, a, b)$.

(iii) Suppose now that there are N firms in the Cournot model of Section 5.1. Extend the code from that chapter to solve numerically for the Nash equilibrium. Check that, for $N = 2$, the code behaves in the same way as `SolveCournotNash`.

(iv) Consider again the two-player case of Section 5.1. However, assume now that Firm 1 moves first, and that its decision is observed by Firm 2 (that is, assume a Stackelberg problem). Adapt the code from Section 5.1 to solve this game.

(v) Let's return to the binary Bayesian game in Section 5.2. Write a function, `BestResponseLoss`, to return the difference between the lefthand side and the righthand side of Equation 5.22. Use this function to create a second function, `SolveBestResponse`, to solve the optimal cutoff value for player i (x_i^*) given the cutoff used by player j (x_j^*).

(vi) Write a function, `BayesianNashLoss`, to do for the model in Section 5.2 what `CournotLoss` did for the model in Section 5.1.

(vii) Write a function, `SolveBayesianNash`, to do for the model in Section 5.2 what `SolveCournotNash` did for the model in Section 5.1.

(viii) Combine `SolveBestResponse` and `SolveBayesianNash` to generate Figure 5.5.

Part III
Dynamics

6 Dynamic Choice on a Finite Horizon

The future ain't want it used to be.

Attributed to Yogi Berra

The steady march of time casts a pall over many economic decisions. The choice to do something today may preclude, restrict, encourage or necessitate certain choices in the future. Searching for work, investing in human capital, installing or removing physical capital, and choosing a partner on the marriage market are all decisions that are often considered from a dynamic point of view.

6.1 Direct Attack

The mathematics of dynamic optimization is often more complex than in the static case. However, the basic idea remains intuitive. In a static context, a consumer acts to equalize the cost-adjusted marginal utility of each unit of consumption. Similarly, when determining behaviour over a number of periods, a decision maker should equalize their discounted marginal utility *at each point in time*. While in a static sense we expect that a consumer should not be able to increase utility by rearranging consumption between goods, in a dynamic sense we would expect that such an improvement cannot be made by rearranging consumption over time. This kind of 'no-arbitrage' type condition has a special name in dynamic optimization: the Euler equation. We will rely on this condition repeatedly.

We start by considering deterministic dynamic problems—that is, problems in which the decision maker faces *no* shocks. In this context, the dynamic problem can be treated as a static problem: a decision maker can optimize perfectly by taking a once-and-for-all decision about their future actions. This is what Adda and Cooper (2003) refer to as 'Direct Attack', and corresponds to Stokey and Lucas (1989)'s 'sequence problem'.

6.1.1 LINEAR FLOW EQUATIONS

Let's start with a simple dynamic problem. Suppose that a household has an endowment, and must decide how much to consume in each of T periods. This is sometimes known as a cake-eating problem on a finite horizon. Assume that the household wants to maximize its discounted utility over these T periods:[1]

$$U = \sum_{t=1}^{T} \beta^{t-1} u(c_t),$$ (6.1)

where β is the household's discount factor.

Let the initial value of the endowment be k_1, which the household apportions over time as it sees fit. In period 1, the household consumes c_1, leaving a maximum consumption of $k_1 - c_1$ for period 2. Generically, this relationship is expressed by the flow equation:

$$k_{t+1} = k_t - c_t.$$ (6.2)

This equation keeps track of capital, our *state variable* (think *state = stock*). The state in any given period depends only upon the state at the beginning of the period, and the decision the individual makes with respect to the choice variable, c. The choice variable, c is also known as the *control variable*.

Equations 6.1 and 6.2, along with the non-negativity constraints $c_t \geq 0$ and $k_t \geq 0$, allow us to completely characterize the household's problem:

$$\max_{\{c_t\}_1^T} \sum_{t=1}^{T} \beta^{t-1} u(c_t)$$ (6.3)

subject to:

$$\sum_{t=1}^{T} c_t + k_{T+1} = k_1$$

$$c_t > 0$$ (6.4)

$$k_t > 0.$$

Here you will notice that we have rearranged the series of flow equations (6.2) into one equation for ease of presentation (and later ease of computation).

This problem looks remarkably similar to the static optimization problems that we have already tackled in earlier chapters. Indeed, if we know the form of $u(c_t)$, the discount factor β, the initial endowment k_1, and the number of periods T, we can solve this problem easily using fmincon. Assume for now

---

[1] Note that we assume additively separability of the utility function in consumption. This assumption is equivalent to assuming that 'the marginal rate of substitution between lunch and dinner is independent of the amount of breakfast', an analogy that Dixit (1990) attributes to Henry Wan.

a log utility function $u(c_t) = \ln(c_t)$, $T = 10$, a discount factor of $\beta = 0.9$, and an initial endowment of $k_1 = 100$. The function FlowUtility returns the scalar V which is the (negative of the) net present utility associated with a consumption stream C, given these parameters:[2]

FlowUtility.m

```
1  function V = FlowUtility(T,Beta,C)
2  %-----------------------------------------------
3  % PURPOSE: calculates the total utility of
4  % consumption assuming an additively separable
5  % utility function and discount rate Beta.
6  %-----------------------------------------------
7  % INPUTS:   C : Tx1 vector of independent variable
8  %           T : scalar time
9  %           Beta : scalar discount rate
10 %-----------------------------------------------
11 % OUTPUT:   V : -utility
12 %-----------------------------------------------
13
14    c   =   C(:,1);
15
16    t   =   [1:1:T];
17    V   =   Beta.^(t-1)*log(c);
18    V   =   -V;
19
20 return
```

Let's now use fmincon to solve for the optimal consumption path. We require strictly positive consumption, c, in each period (as log(0) is undefined). We achieve this in MATLAB by setting a lower bound of eps, which equals 2^{-52}, for each of the 10 periods.[3] An upper-bound is also defined, as consumption can never exceed 100 (the full amount of the endowment) in any period. We need an initial starting point (guess). For simplicity, this is defined as equal

---

[2] The final line of the script, converts the (positive) utility V into a negative value, as although we are interested in *maximizing* utility, fmincon is a *minimization* function. Also, on line 14 we ensure that consumption is a 1 × T row vector, which allows us to recycle the script later in this chapter.

[3] Type help eps. You will see that eps is formally defined as 'the distance from 1.0 to the next largest double-precision number'. To see intuitively how eps works, try the following two logical tests:

```
» 1 == (1+ eps)
ans =
0
» 1 == (1+ eps/2)
ans =
1
```

consumption in all periods. Finally, we set up the flow constraint that total consumption must not exceed the full endowment k1. We do this using the vector A, defining that $A \cdot c \leq k_1$.

```
>> Beta    = 0.9;
>> T       = 10;
>> K1      = 100;
>> lb      = eps*ones(10,1);
>> ub      = 100*ones(10,1);
>> guess   = 10*ones(10,1);
>> A       = ones(1,10);
>> opt     = optimset('TolFun', 1E-20, 'TolX',...
             1E-20, 'algorithm', 'sqp');
>> c       = fmincon(@(C) FlowUtility(T,Beta,C),...
             guess, A, K1, [], [], lb,...
             ub, [], opt)

c =

    15.3534
    13.8181
    12.4363
    11.1926
    10.0734
     9.0660
     8.1594
     7.3435
     6.6091
     5.9842
```

The vector c gives optimal consumption in each period. As expected (given that $\beta < 1$), the consumption profile is downward sloping. Note that the household optimally consumes the entirety of their endowment k_1:

```
>> sum(c)

ans =

    100
```

In this example, we have made a number of assumptions about the household's preferences—namely, the particular value of their discount rate, and the log functional form for household utility. In Exercises (i)–(ii) at the end of this chapter, we ask you to explore the consequences of these assumptions for the profile of optimal consumption.

6.1.2 NON-LINEAR FLOW EQUATIONS

In Section 6.1, we were able to reduce the 'dynamism' of the problem by rearranging our flow constraints (Equation 6.2) into one simple constraint (Equation 6.4). This relied on a number of strong assumptions that might be thought rather artificial. In this section, we will relax these, to consider a more realistic example in which the flow constraints cannot be rearranged into such a nice linear format.

Imagine now that our household is both a producer and a consumer. For example, we could think of a household microenterprise, whose consumption decisions in one period affect future production. Specifically, we will assume that the household extracts c_t from their capital stock at the beginning of period t leaving $k_t - c_t$ as an input to the (strictly concave) production function, f. The flow equation now takes the form:

$$k_{t+1} = f(k_t - c_t, \theta), \tag{6.5}$$

where θ is a time-invariant technology parameter.

The household's maximization problem now becomes:

$$\max_{\{c_t\}_1^T} \sum_{t=1}^{T} \beta^{t-1} u(c_t) \tag{6.6}$$

subject to:

$$k_{t+1} = f(k_t - c_t, \theta) \tag{6.7}$$

$$c_t > 0 \tag{6.8}$$

$$k_t > 0. \tag{6.9}$$

As opposed to the problem in the previous section, we can no longer specify our maximization problem with one simple linear flow constraint (as was the case with the condition $\mathbf{A} \cdot \mathbf{c} = k_1$). This is because Equation 6.8 is non-linear. This non-linearity implies that high consumption in early periods not only runs down the stock of k but also affects the household's ability to produce more k in the future.

Let's assume that the production function is Cobb-Douglas:

$$k_{t+1} = f(k_t - c_t, \theta) = \theta(k_t - c_t)^\alpha. \tag{6.10}$$

To solve this dynamic maximization directly, we can form a series of T non-linear constraints. As we saw in the previous chapter, fmincon allows for non-linear constraints to be passed as a function. FlowConstraint defines these for our current problem.

FlowConstraint.m

```
1  function [d,deq] = ...
2  FlowConstraint(CK,T,K1,Theta,Alpha)
3  %-----------------------------------------------
4  % PURPOSE: sets up the system of constraints
5  % k_{t+1}=\theta (k_t-c_t)^\alpha
6  %-----------------------------------------------
7  % INPUTS: CK    : Tx2 matrix of C and K values
8  %                   at each t
9  %          T     : scalar time
10 %          K1    : stock of K at start of t = 1
11 %          Theta : Cobb-Douglas parameter
12 %          Alpha : Cobb-Douglas parameter
13 %-----------------------------------------------
14 % OUTPUT:  d   : inequality constraint vector
15 %          deq : equality constraint vector
16 %-----------------------------------------------
17
18    cap   =   CK(:,2);
19    c     =   CK(:,1);
20    k     =   [K1; cap];
21
22    for t = 1:T
23          deq(t) = k(t+1)-Theta* (k(t)-c(t))^Alpha;
24    end
25    d   =   [];
26
27 return
```

It is important to note that any non-linear constraints function passed to fmincon *must* return two outputs; in FlowConstraint we call these d and deq.[4] These correspond to the non-linear equalities and non-linear inequalities. Given that Equation 6.5 is a system of T *equality* constraints, we just define an empty vector for the inequality constraints d. Before moving on to solve this all using fmincon, we suggest that you experiment with this function, perhaps starting with Exercise (iii) at the end of this chapter.

Let's now run our code . . .

```
>> Beta = 0.9;
>> T = 10;
```

---

[4] In the previous chapter, we used the standard fmincon terminology and called these non-linear constraints c and ceq—but, in this context, it is more natural for c to refer to consumption.

```
>> K1 = 100;
>> Theta = 1.2;
>> Alpha = 0.98;
>> lb     = zeros(10,2);
>> ub     = 100*ones(10,2);
>> guess = [10*ones(10,1), [90:-10:0]'];
>> opt   = optimset('TolFun', 1E-10, 'TolX', 1E-10,...
             'algorithm','sqp',...
             'MaxFunEvals', 100000,'MaxIter', 2000);
>> result = fmincon(@(CK) FlowUtility(T,Beta,CK),...
             guess,[],[],[],[],lb,ub, @(CK)...
             FlowConstraint(CK,T,K1,Theta,Alpha),opt)

result =

     16.5011    91.7125
     15.9856    83.3386
     15.5165    74.8041
     15.0944    66.0245
     14.7213    56.9014
     14.4010    47.3159
     14.1408    37.1173
     13.9543    26.1024
     13.8695    13.9624
     13.9624     0.0000
```

The optimization returns a two-column matrix with $T = 10$ rows. This is our result for CK: consumption (c_t) and remaining capital ($k_t - c_t$) in each period. Let's graph these variables as in Figure 6.1. We can see that all capital is consumed and that the household/firm has an approximately downward sloping consumption profile.

```
>> plot(1:T, result(:,1), '--r', 1:T, result(:,2),...
     'linewidth', 2)
>> xlabel('Time', 'FontSize', 14)
>> ylabel('C_t,k_t', 'FontSize', 14)
>> legend('Consumption', 'Capital Remaining',...
     'Location', 'NorthEast')
>> title({'Firm Consumption and Investment',...
       '\beta=0.9, \theta=1.2, \alpha=0.98'},...
     'FontSize', 16)
>> print -depsc DynamicBehaviour
```

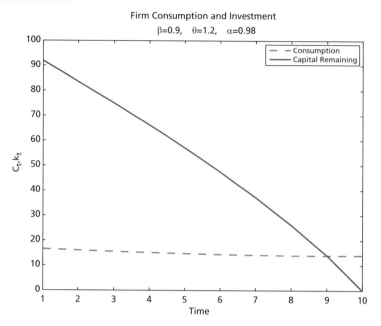

Figure 6.1 Dynamic Behaviour of a Household Firm

6.2 **Dynamic Programming**

Direct Attack is a very useful way of solving deterministic dynamic problems. Sadly, the real world is rarely so simple—typically, we must plan for the future without knowing exactly what the future will look like. We need a method to allow us to model dynamic decision making under uncertainty. This is the beauty of dynamic programming,

The idea behind dynamic programming is simple: rather than solving a complex optimization problem that considers all T periods all at once, we can optimize one period at a time. To do this we must form a single summary statistic for 'the future': the value of all remaining capital, assuming that this capital is used optimally. For this reason, dynamic problems on a finite horizon can be solved by backward induction: we first solve for the final period, then for the penultimate which depends upon the final, and continue this until we arrive at the first period.

To implement this in practice, we must introduce the concept of the value function. To illustrate, consider again the household cake-eating problem. In this context, the value function summarizes the value to the household of a given amount of capital, assuming that this capital will be used optimally in future periods. Start with the final period. We know that the value of any capital which remains beyond the final period is zero, given that it spoils:

$$V_{T+1}(k_{T+1}) = 0 \tag{6.11}$$

This gives us a place to start for our backward induction. When making its decision in period T, the household solves:

$$V_T(k_T) = \max_{c_T \in (0,k_T]} \{u(c_T) + \beta V_{T+1}(k_{T+1})\}, \tag{6.12}$$

where $k_{T+1} = k_T - c_T$. Given that we already have our terminal condition from Equation 6.11, Equation 6.12 can be solved for any k_t.[5] This solution gives the total value to the household of entering period T with some amount k_T, and then behaving optimally. This in turn allows us to consider the decision in period $T - 1$:

$$V_{T-1}(k_{T-1}) = \max_{c_{T-1} \in (0,k_{T-1}]} \{u(c_{T-1}) + \beta V_T(k_T)\}, \tag{6.13}$$

exactly analogous to Equation 6.12. Here we start to see the process of backwards induction. Once we solve Equation 6.13, we can then move on to solve for $V_{T-2}(k_{T-2})$, and continue until reaching $V_1(k_1)$.

Time for some MATLAB...

BackwardsInduc.m

```matlab
%----- (1) Prompt user to input parameters
Beta        =   input('Input Beta:');
T           =   input('Input time:');
K1          =   input('Input initial capital:');
grid        =   input('Input fineness of grid:');

K       =   0:grid:K1;
V       =   [NaN(length(K),T), zeros(length(K),1)];

%----- (2) Loop over possible values of k_{t} and
%             k_{t+1}---
aux     =   NaN(length(K),length(K),T);
for t   =   T:-1:1
  for inK   =   1:length(K)
    for outK    =   1:(inK)
      c             = K(inK)-K(outK);
      aux(inK,outK,t)   =   log(c)+Beta*V(outK,t+1);
    end
  end
  V(:,t)=max(aux(:,:,t),[],2);
end
```

[5] The solution for any value of k will be $k_T = c_T$. Why?

```
22
23  %----- (3) Calculate optimal results going forward
24  vf    = NaN(T,1);
25  kap   = [K1; NaN(T,1)];
26  con   = NaN(T,1);
27
28  for t=1:T
29    vf(t)    =V(find(K==kap(t)),t);
30    kap(t+1) =K(find(aux(find(K==kap(t)),:,t)==vf(t)));
31    con(t)   =kap(t)-kap(t+1);
32  end
33
34  %----- (4) Display results --------------------
35  [kap([1:T],:),con]
36  subplot(2,1,1)
37  plot([1:1:T],[con, kap([2:T+1],:)],'LineWidth',2)
38  ylabel('Consumption, Capital', 'FontSize', 12)
39  xlabel('Time', 'FontSize', 12)
40  legend('Consumption', 'Capital')
41
42  subplot(2,1,2)
43  plot([1:1:T], vf, 'Color', 'red', 'LineWidth', 2)
44  ylabel('Value Function', 'FontSize', 12)
45  xlabel('Time', 'FontSize', 12)
```

To understand what BackwardsInduc.m is doing, let's start with section 2 of the code. Let's ignore for the time being the outer loop (which starts on line 13), and focus on the inner two loops. Here we define the matrix aux, which looks very similar to Equation 6.12 and Equation 6.13. aux tells us, for each k_t, all possible exit values of k_{t+1}. For example, if we enter a given period with 20 units of k_t, the household could choose to consume 20 now and 0 in the future, 19 now and 1 in the future, 18 now and 2 in the future, and so forth.

Once we have calculated this matrix for all possible input and output capital values, we can calculate the optimal decision for each possible input capital. This is what we do on line 20. The matrix V tells us the best possible behaviour for any given k_t—for example, were we to arrive to a given period with $k_t = 20$, we may find that the optimal choice is to consume 5 now, and save 15 for the future (and so can discard the other 20 possible combinations).

You may wonder why we bother doing this for each possible input capital value. For example, why is it important to know what the household would do if it were to arrive with 20 units, if in reality it arrives and has 19 units of capital? The reason is simple: we will not know what decisions the household faces in each period until we completely solve the model, and to be able to solve the model we must know what the value function looks like in future periods. We call the vector V the value *function* because it is a solution for all possible values of k.

This brings us to the heart of dynamic programming (at least when considering problems with a finite horizon). When solving, we must first iterate backward; only then can we iterate forward to obtain the objective function. This is why you see two loops involving time (those starting at line 13 and at line 28) in the above code. The first of these loops calculates the value function starting in period T and counting backwards to the first period. The second loop determines how much capital to consume in each period, starting from period 1 and ending in period T. We must start in period 1 because this is the only period where we know with certainty what the beginning capital will be.

Let's now run our code to see what happens. To test how this compares to a direct attack, we will input the same values as earlier:

```
>> BackwardsInduc
Input Beta:0.9
Input time:10
Input initial capital:100
Input fineness of grid:0.25

ans =

   100.0000     15.5000
    84.5000     13.7500
    70.7500     12.5000
    58.2500     11.2500
    47.0000     10.0000
    37.0000      9.0000
    28.0000      8.2500
    19.7500      7.2500
    12.5000      6.5000
     6.0000      6.0000
```

Figure 6.2 plots the solution for total remaining capital and consumption in each period, as well as the value function itself. Compare the consumption values—the second column of the above output—with those calculated from the direct attack in Section 6.1.1. Notice that we have lost some precision. This is because we have had to discretize the possible values of capital, k_t and k_{t+1}. In section 1 of the code, we define the possible values of capital as K = 0:grid:K1;. This allows the household to choose any possible values between 0 and the full amount K1, in steps of size grid. You should check how

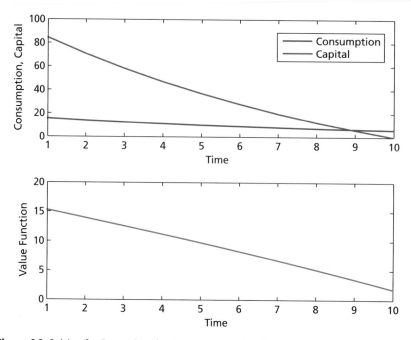

Figure 6.2 Solving for Dynamic Behaviour Using a Value Function

this solution changes as we define finer and finer grids over which to search.[6] Before we move on and discuss how to save time with these sorts of grid search options, we suggest that you experiment further with this code. Exercise (v) at the end of the chapter is particularly relevant, and lets you build your own value function for a problem we have tackled earlier.

6.3 **Memoization**

We have just seen that the accuracy of our solution depends upon the fineness of the search grid. Of course, if we want a highly accurate solution, we could just specify a very fine grid, such as an increment of 0.001. However, this is very demanding on the processing capacity of most personal computers.

The main bottleneck in this code is the calculation of each possible capital pair combination—that is the `aux` matrix in the last example. For example, if we specify a `grid = 0.1`, we see that:

---

[6] For general interest, with a grid size of 0.05 we find that `Result(:,2)'` = 15.3500 13.8000 12.4500 11.2000 10.1000 9.0500 8.1500 7.3500 6.6000 5.9500.

```
>> size(aux)

ans =

    1001 1001 10
```

Were we patient (or brave?) enough to try with a grid size of 0.001, we would be dealing with a matrix with 100 billion individual elements. Perhaps we would be willing to accept waiting a reasonable time to solve this one problem very accurately—but it is unlikely that we could afford such a luxury if we were resolving this for many households (rather than one), or if we were dealing with more than one state variable. Such a situation is well known in dynamic programming, and was labelled the 'Curse of Dimensionality' by Bellman (1957).[7]

Memoization is one response to the Curse of Dimensionality in dynamic programming. This refers to the process of 'remembering', rather than re-computing, for use in subsequent analysis. As Michie (1968) suggests, 'It would be useful if computers could learn from experience and thus automatically improve the efficiency of their own programs during execution. A simple but effective rote-learning facility can be provided within the framework of a suitable programming language.'

To understand the appeal of memoization, return to step 2 of the earlier code. Here we entirely solve the dynamic problem for any possible starting and finishing value of capital in each period. Only once we have completely solved this problem can we compute what the household does when starting with a capital value of $k_1 = 100$. Now imagine that we want to solve for the behaviour of a household with $k_1 = 50$. We already have our memoized solution from step 2, so do not need to do any further backward induction. Similarly, if our household unexpectedly receives an additional amount of capital in between periods 3 and 4, we simply follow our memoized solution, avoiding the main cost of calculation. We will see the importance and flexibility of such a situation below, where we consider stochastic dynamic programming.

6.4 **Stochastic Dynamic Programming**

Time for some shocks! Returning to our household microenterprise, let's imagine that the evolution of capital is subject to a stochastic shock. The transition equation now becomes:

$$k_{t+1} = f(k_t - c_t, \theta, \varepsilon_{t+1}) = \theta(k_t - c_t)^\alpha + \varepsilon_{t+1}. \qquad (6.14)$$

While the decision maker does not know the precise value of ε_{t+1} at time t, we assume that she does know its distribution.

---

[7] We will meet a similar curse in a different context in Chapter 9.

Rather than deciding between consumption now and consumption in the future, the household must now choose between consumption now and *expected* future consumption. This requires a rewriting of the dynamic programming problem:

$$V_t(k_t) = \max_{c_t \in (0,k_t]} \{u(c_t) + \beta \mathbb{E}[V_{t+1}(k_{t+1})]\}, \tag{6.15}$$

where k_{t+1} is given as in Equation 6.14, and the expectation is taken over the distribution of ε_{t+1}.

To illustrate, let's assume a very simple distribution for the ε term. Specifically, we assume that ε takes two possible states: low ($\varepsilon = -2$) and high ($\varepsilon = 2$). Let each state occur with a probability of 0.5.[8] The stochastic portion of this problem is then characterized by two vectors: a vector of shocks ($\boldsymbol{\varepsilon} = [-2, 2]$) and a vector of probabilities ($\boldsymbol{\pi} = [0.5, 0.5]$). V_{t+1} now depends on future capital and the realization of future shocks.

The code below computes the value function at each time period, for each possible (optimal) capital–consumption pair. nextKl refers to the future capital stock (k_{t+1}) associated with a low realization of the shock (i.e. $\varepsilon_{t+1} = -2$), and nextKh for a high realization.[9] Line 26 is MATLAB speak for 'replace nextKl with zero if nextKl is less than zero'. This is a technological requirement: in this model, capital can never be negative.

The expected future value is given by EnextV. To calculate this, we find the value function associated with the low shock and with the high shock; we then weight by their respective probabilities. To ensure that the predicted k_{t+1} lies in the discretized state space for capital, nextK is rounded to the closest value in our capital grid. In dynamic problems with continuous state variables, discretization steps such as this are necessary for computation.

FiniteStochastic.m

```
1   clear; clc;
2
3   %----- (1) Setup parameters -------------------
4   e      =   [-2 2];
5   PI     =   [0.5 0.5];
6   Beta   =   0.9;    Theta = 1.2;    Alpha = 0.98;
7   K1     =   100;
8   grid   =   0.1;
```

[8] Stachurski (2009) provides an excellent overview of modelling stochastic processes with dependence.

[9] Where these are calculated in the code, rather than typing nextKj = theta*(K(inK) - c)^alpha + e(j), we could have just typed nextKj = K(outK) + epsilon(j). The logic is exactly the same: we add the shock that occurs *after* the current period terminates to the capital available at the end of this period.

```
9   T          =   10;

10

11  K          =   0:grid:K1+max(e);
12  V          =   [NaN(length(K),T), zeros(length(K),1)];
13  aux        =   NaN(length(K),length(K),T);

14

15

16  %----- (2) Loop over possible values of c, k and
17  %              epsilon -----
18  for t = T:-1:1
19  fprintf('Currently in period %d\n', t)

20

21    for inK = 1:length(0:grid:K1)
22      for outK = 1:inK
23        c        = K(inK) - (K(outK)/Theta)^(1/Alpha);
24        nextKl   = Theta*(K(inK) - c)^Alpha + e(1);
25        nextKh   = Theta*(K(inK) - c)^Alpha + e(2);
26        nextKl(nextKl<0) = 0;

27

28        EnextV = PI(1) * V((round(nextKl/grid)+1),...
29        t+1) + PI(2) * V((round(nextKh/grid)+1),t+1);
30        aux(inK,outK,t)  =  log(c) + Beta*EnextV;
31      end
32    end
33    V(:,t)=max(aux(:,:,t),[],2);
34  end
```

This script calculates the value function at each of ten time periods, for each possible (optimal) capital–consumption pair. We can save the results (that is, 'memoize' the function), and examine the optimal consumption paths for particular realizations of the shock term.

Unlike the deterministic model of Section 6.1, the stochastic model does not have a single optimal solution—rather, the optimal consumption path depends upon the particular realization of ε. The code below draws values of ε for 100 different households. This shows how optimal consumption depends on the stochastic component of the model.

SimulateStochastic.m

```
1   %----- (1) Set-up parameters, simulate shocks ---
2   people      =   100;

3

4   epsilon     =   randi(2,people,T+1);
```

```
 5  epsilon(epsilon==1)    =    -2;
 6
 7  vf           =     NaN(people,T);
 8  kap          =     [K1*ones(people,1) NaN(people,T)];
 9  con          =     NaN(people,T);
10
11  %----- (2) Determine consumption based on
12  %                simulated shocks ----
13  for p=1:people
14    for t=1:T
15     position     =    round(kap(p,t)/grid+1);
16     vf(p,t)      =    V(position,t);
17     kap(p,t+1)   =    K(find(aux(position,:,t) ...
18                       ==vf(p,t)));
19     con(p,t)  =   kap(p,t)-(kap(p,t+1)/Theta)^...
20                    (1/Alpha);
21     kap(p,t+1)   =    kap(p,t+1)+epsilon(p,t+1);
22    end
23  end
24
25  %----- (3) Output ----------------------------
26  plot([1:1:T], con)
27  ylabel('Consumption', 'FontSize', 12)
28  xlabel('Time', 'FontSize', 12)
29  title('Simulated Consumption Paths',...
30        'FontSize', 16)
31
32  figure(2)
33  hist(sum(con,2))
34  title('Lifetime Consumption', 'FontSize', 16)
```

The loop starting on line 13 calculates the optimal decision for each household: consumption (con), capital (kap) and the resulting value function (vf) at each time period. Notice that we use the memoized matrix V that we calculated earlier. This illustrates a fundamental aspect of most stochastic dynamic programming problems on a finite horizon: we iterate *backwards* to solve the model, then *iterate* forwards to simulate it.

Figure 6.3 shows the optimal consumption paths for our 100 simulated households. In period 1 (where each household has the same capital endowment) everyone acts in precisely the same way. However, in the following periods, optimal behaviour diverges because of differences in the realizations of ε (and hence k).

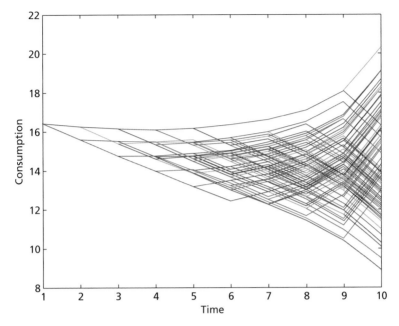

Figure 6.3 Simulated Consumption in a Stochastic Model

6.5 **Estimating Finite Horizon Models**

So far, we have been simulating models under the assumption that we know the relevant underlying structural parameters: β, α, and θ. Let's refer to this group of parameters as $\boldsymbol{\Omega}$:

$$\boldsymbol{\Omega} = (\beta, \theta, \alpha). \tag{6.16}$$

These simulations are useful to predict behaviour under a certain parametric representation of the world—but often our interest is in inverting this problem. Rather than assume that we know $\boldsymbol{\Omega}$, suppose that we have data on household behaviour and are interested in estimating these underlying parameters.

6.5.1 ESTIMATING BY GMM

In Chapter 3, we used GMM to estimate a linear regression. The dynamic models introduced in this and the previous chapter are entirely amenable to estimation in a similar way. Perhaps the trickiest part of this process is to motivate and justify specific moment conditions.

To fix ideas, let's return to the household microenterprise example of Section 6.4. Method of Moments estimation starts with an assumption that *something* in the population is, on average, equal to zero. Let's assume that we have reason to believe that the expected value of the stochastic error term is zero in each period:

$$\mathbb{E}[\varepsilon_t] = 0 \ \forall \ t. \tag{6.17}$$

Notice that this is just an assumption about *one moment* (the mean of ε); we do not assume anything else about its distribution.[10]

Recall that the maximization problem for our household microenterprise is:

$$\max_{\{c_t\}_{t=1}^T} \sum_{t=1}^T \beta^{t-1} u(c_t) \quad \text{subject to} \quad k_{t+1} = f(k_t, c_t) + \varepsilon_{t+1}. \tag{6.18}$$

By combining Equations 6.17 and 6.18, we can write moment conditions. The first set of conditions come from simply rearranging the capital flow equation, expressing in terms of ε, and taking expectations of both sides:

$$\mathbb{E}[k_{t+1} - f(k_t, c_t)] = 0. \tag{6.19}$$

The second set of conditions comes from the Euler equations. These require that the marginal rate of substitution of consumption between periods is equal to the marginal return on saved capital:

$$\mathbb{E}\left[\frac{u'(c_t)}{\beta u'(c_{t+1})} - f'(k_t)\right] = 0. \tag{6.20}$$

With our assumed functional forms, these moment conditions become:

$$\mathbb{E}[k_{t+1} - \theta(k_t - c_t)^\alpha] = 0 \tag{6.21}$$

$$\mathbb{E}\left[\frac{c_{t+1}}{\beta c_t} - \alpha\theta(k_t - c_t)^{\alpha-1}\right] = 0. \tag{6.22}$$

With population moment conditions in hand, we can now fit the sample analogue of these moments using the data we simulated earlier (see `SimulateStochastic.m`). In dynamic models, the discount factor β is typically not identified without placing strong restrictions on other primitives in the model.[11] Therefore, we will assume a plausible value for β, plug this into our estimation, and estimate the remaining parameters.

---

[10] This is less demanding than assumptions we have used earlier in the book—for example, when estimating by Maximum Likelihood in Chapter 4, we made an assumption about the full distribution of ε. This is perhaps one of the most attractive features of moment based estimation. We return to this point at the end of this section.

[11] We will not go into this here—however, an exposition can be found in Rust (1994a,b).

Let's write these moment conditions in MATLAB:

DynamicMoments.m

```
 1  function Q=DynamicMoments(ct,ctp,kt,ktp,params)
 2  %------------------------------------------------
 3  % PURPOSE: returns the quadratic distance based
 4  % on dynamic model moments and specified values
 5  % of alpha and theta
 6  %------------------------------------------------
 7  % INPUTS:    ct      : Nx1 vector of C_t
 8  %            ctp     : Nx1 vector of C_{t+1}
 9  %            kt      : Nx1 vector of K_t
10  %            ktp     : Nx1 vector of K_(t+1)
11  %            params  : [Alpha; Theta]
12  %------------------------------------------------
13  % OUTPUT:    Q : value of moment conditions
14  %------------------------------------------------
15
16  Alpha    =  params(1);
17  Theta    =  params(2);
18  Beta     =  0.9;
19  k        =  size(params,2);
20
21  %----- (1) Form moments -----------------------
22  m1  = mean([ktp - Theta*(kt - ct).^Alpha]);
23  m2  = mean([(ctp./(Beta*ct))-Alpha*Theta*...
24        (kt-ct).^(Alpha-1)]);
25
26  %----- (2) Create weight matrix and quadratic
27  %           distance -------
28  W        = eye(k);
29  Q        = [m1 m2]*W*[m1 m2]';
30  return
```

Notice that the moments m1 and m2 are based on the functional form outlined above and that we have assumed $\beta = 0.9$. Notice also that we have only formed two moments. In Equation 6.17, we have assumed that $\varepsilon = 0$ in each of our T time periods. This means that we could create an overidentified system of moments; we invite you to do this in Exercise (vi).

To estimate, we will use data from just one time period. We will start by simulating our data, and then use fminunc—MATLAB's unconstrained minimization routine—to minimize our objective function:

```
>> FiniteStochastic;
>> SimulateStochastic;
>> opt = optimset('TolFun', 1E-20, 'TolX', 1E-20);
>> [Param, Q] = fminunc(@(p) DynamicMoments...
                   (con(:,4),con(:,5),kap(:,4),...
                    kap(:,5),p),[1, 1], opt)


Param =

    0.9832    1.1895


Q =

    2.7558e-08
```

Our estimates of 0.9832 and 1.1895 are close to the true population values of 0.98 and 1.20. In Exercises (vi) and (vii), we encourage you to improve these estimates by using a larger set of moment conditions. We encourage you also to experiment with the code to see how it performs under alternative (and less serendipitous) assumptions regarding functional form, discount rate, and so forth.

6.6 Review and Exercises

Table 6.1 Chapter 6 Commands

Command	Brief Description
subplot	Display multiple graphs on one output
input	Prompt user input from the keyboard
find	Find location(s) of exact coincidence in a matrix
print	Print to disk the item currently in graphical memory
clc	Clear results screen
figure	Output various figures in a MATLAB script

This chapter has introduced time into economic models. This is a major area of current research and application, as a glance through the *Journal of Economic Dynamics and Control* will attest. For those interested in continuing with this line of research, useful places to start for more extensive (textbook) analyses are Stokey and Lucas (1989), Adda and Cooper (2003), or chapters of Dixit (1990) and Acemoglu (2008). We also, of course, encourage you to read Chapter 7 of this book. Beyond this, the following topics and resources may serve as useful stepping stones along your path.

First, in solving for the expected value function, a number of more complex tools can be called upon. Perhaps most importantly, numerical integration is often used, integrating over the value of the stochastic proportion of the model. A useful starting point for numerical integration is Judd's (1998) text book. In MATLAB try typing `help integrate` or `lookfor integral`. In numerically computing integrals, quadrature methods are typically used. Gauss-Legendre or Guass-Hermite quadrature are perhaps the most well-known examples.

Second, we can incorporate more complex stochastic elements. In the stochastic model we have considered in this chapter, the stochastic elements have been quite simple: we have assumed that the uncertainty can be characterized by a known distribution, which is invariant over time. More flexible structures can be built using Markov Chain structures, where the value of the shock can depend upon the value of the shock in previous periods. Typically, this involves defining a transition probability matrix along with a set of states between which these transitions occur. There are many resources which can be consulted for reading. Those particularly related to dynamic programming include Adda and Cooper (2003), Stachurski (2009), and Ljungqvist and Sargent (2000).

Third, GMM is not the only way to estimate these types of models. Other options to consider include Simulated Method of Moments, Maximum Likelihood, and Maximum Simulated Likelihood. We cannot suggest a much better place to start than Eisenhauer et al. (2014), who discuss precisely this issue. Similarly, you should consult Keane et al.'s (2013) handbook chapter—with its excellent closing remarks on 'how credible are DCDP models'. For a more general discussion of inference in economic (often dynamic) models, see Wolpin (2013).

Finally, after reviewing Eisenhauer et al. (2014), it is worth examining other empirical papers that estimate dynamic models, to see the modelling and estimation choices made. Useful examples include Wolpin (1984), Keane and Wolpin (1997), Adda and Cooper (2000), and Todd and Wolpin (2006).

EXERCISES

(i) In the function `FlowUtility.m`, we have experimented with optimal consumption patterns based on a given discount rate `Beta`. How does the optimal consumption path vary with β? Write your own script to produce a graph (perhaps like either of the following), to show optimal consumption over a range of βs.

(ii) How do our results depend upon the utility function we specify? Can you generalize the function `FlowUtility.m` to work for various different utility functions? For example, can you make an additional argument (modifying the help file of course!) so that the function now also allows for isoelastic utility, and exponential utility?

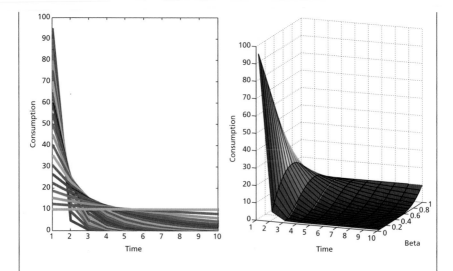

(iii) Try experimenting with the `FlowConstraint.m` function by passing it the entire set of arguments, to see what it returns in each case. Try situations in which deq equals zero (perhaps where $\alpha = 1$ and $\theta = 1$) and where deq $\neq 0$. *A useful hint*: in order to see both outputs (d and deq) you must request these explicitly from MATLAB.

(iv) Our models in this chapter, based on the utility function in Equation 6.1, have assumed a constant discount rate β. What happens, however, if we assume that our agents are time inconsistent? Specifically, what happens if they exhibit quasi-hyperbolic (or β, δ) discounting? Experiment with values of β and δ (perhaps start with $\beta = 0.7, \delta = 0.96$), and see if this is inline with your intuition (As a reminder, quasi-hyperbolic discounting implies a discount factor of $[1, \beta\delta, \beta\delta^2, \beta\delta^3, \ldots]$; for further details on these behaviours, see Laibson (1997)).

(v) In BackwardsInduc.m we have solved for a flow equation of the form $k_{t+1} = k_t - c_t$. Modify the code to solve for $k_{t+1} = \theta(k_t - c_t)^\alpha$ (and confirm that this is correct by referring to the results from section 6.1.1). *A useful hint*: this can be incorporated into BackwardsInduc.m with two fairly minor changes. Concentrate only on the formulas which define consumption (c and con).

(vi) When estimating our dynamic model using GMM, we calculated $\hat{\alpha}$ and $\hat{\theta}$ by using two moment conditions based upon ε_5. Generalize this GMM estimation so that rather than using two conditions you use all the (T-1)*2 moments available from ε_t.

(vii) In Chapter 3, you may remember that we asked you to generalize the one-step GMM code provided there, to estimate the more efficient two-step procedure. Do the same here, generating an optimal weight matrix $\widehat{W}$ based on the first-step estimates. (*A useful hint*: You can find the formula for $\widehat{W}$ in the exercises in Chapter 3. Alternatively, this is identical to using the covariance matrix of the moments in the first step.)

7 Dynamic Choice on an Infinite Horizon

Bigger than the biggest thing ever and then some. Much bigger than that in fact, really amazingly immense, a totally stunning size, real 'wow, that's big', time. Infinity is just so big that by comparison, bigness itself looks really titchy. Gigantic multiplied by colossal multiplied by staggeringly huge is the sort of concept we're trying to get across here.

Douglas Adams, *The Hitchkiker's Guide to the Galaxy*[*]

Dynamic programming on a finite horizon relies upon backward induction. In a sense, the final period acts as an anchor from which we can calculate the value function—and, as a result, optimal decisions—in all time periods.

In this chapter, we cast that anchor adrift. There are many situations in which decisions do not necessarily have a known final period. For example, it seems unlikely that many firms would plan to close at a defined point in the future. Indeed, even in the case of individuals deciding over their lifetime, the terminal point is uncertain. Problems of this type require another class of solution. Fortunately, such a situation can still be motivated by the ideas of Bellman (1957), which we discussed in Chapter 6.

Let's return to the household enterprise's decision, which was discussed in Chapter 6. In this simple model, the household must decide on how much to consume now and how much to invest in the future. This problem has a long history in economics, being described by Ramsey as early as 1928 (although applied to countries rather than firms; some examples for firms include Bond and Söderbom (2005) and Fafchamps et al. (2014)). The objective is to maximize total discounted utility, subject to the flow equation for capital. Here, let's assume log utility and Cobb-Douglas production, giving the specific optimization problem below:

$$\max_{\{c_t\}_{t=1}^{\infty}, \{k_t\}_{t=2}^{\infty}} \sum_{t=1}^{\infty} \beta^{t-1} \ln(c_t) \qquad (7.1)$$

[*] Excerpt from *The Restaurant at the End of the Universe* by Douglas Adams, 1980, by Serious Productions Ltd. Used by permission of Harmony Books, an imprint of the Crown Publishing Group, a division of Penguin Random House LLC. All rights reserved. Any third party use of this material, outside of this publication, is prohibited. Interested parties must apply directly to Penguin Random House LLC for permission.

subject to:

$$k_{t+1} = \theta k_t^\alpha - c_t \tag{7.2}$$

$$c_t \geq 0 \tag{7.3}$$

$$k_t \geq 0. \tag{7.4}$$

You will notice that rather than the household enterprise maximizing over T periods, they now maximize over an infinite horizon. Typically, finding an analytical solution to dynamic problems of this type is impossible or very difficult—meaning that we *must* revert to numerical tools like MATLAB. However, (luckily for us!) our functional form assumptions allow us to solve the model analytically, providing a useful benchmark against which to compare our results.

7.1 Value Function Iteration

In Chapter 6, we introduced Bellman's functional equation (see Equation 6.12). The basic idea is reasonably simple. We take a dynamic optimization problem and break it into two periods: now (the current period), and the future (all future periods). The value of the future is represented by a 'functional equation', V.

In Chapter 6, we gave the Bellman equation in the following form:

$$V_t(k_t) = \max_{c_t \in (0, k_t)} \{u(c_t) + \beta V_{t+1}(k_{t+1})\}, \tag{7.5}$$

The subscript t has a central role in the above expression. Of course, this should not surprise us—the previous chapter was about finite problems, in which agents may optimally change their behaviour as the terminal period approaches. For this reason, the decision problem must be indexed by t. We call this problem *non-stationary*.

In this chapter, however, we consider problems on an infinite horizon. Without a terminal period to anchor us, we must have a value function that does not depend on time:

$$V(k) = \max_{c, \tilde{k}} \left\{ u(c) + \beta V(\tilde{k}) \right\}. \tag{7.6}$$

In Equation 7.6, k refers to capital in the current period and $\tilde{k}$ refers to capital in the following period. Notice that the value function itself does not depend on time. We can, therefore, describe this problem as *stationary*. Substituting from Equation 7.2 to 7.4, we obtain:

$$V(k) = \max_{c,\tilde{k}} \left\{ \ln(\theta k^\alpha - \tilde{k}) + \beta V(\tilde{k}) \right\}. \qquad (7.7)$$

We can solve Equation 7.7 by iteration. Essentially, this 'value function itera-tion' is searching for a fixed point, at which point we have numerically solved the value function. Start with an arbitrary initial function $V(\tilde{k}) = 0$. Use Equa-tion 7.7 to update to a new value function (i.e. the left-hand side of Equation 7.7). If we then plug this new value function back into the right-hand side of the equation, we will once again get a new value function, which we, yet again, substitute into the right-hand side of Equation 7.7. We can continue this process *ad infinitum*—or rather until the value function on the left-hand side is equal to the right-hand side of Equation 7.7. At this point Equation 7.7 is solved, as we have found the best possible way to allocate capital between consuming now and saving for consumption in the future. In compact form, we can write this process as:

$$\Gamma V(k) = \max_{\tilde{k}} \left\{ \ln(\theta k^\alpha - \tilde{k}) + \beta V(\tilde{k}) \right\} \qquad \forall\, k, \qquad (7.8)$$

where Γ is an operator representing this process of iteration on the value function until $\Gamma V(k) = V(k)$. In the Appendix to this chapter, we show how value function iteration works analytically in this case. If you require further convincing of the mechanics, this is a worthwhile example to work through. As derived in this Appendix, we show that the earlier Bellman equation (Equation 7.7) can be satisfied by a value function of the form:

$$V(k) = \frac{\alpha}{1 - \beta\alpha} \ln k + F \qquad (7.9)$$

where F just represents a constant, and that this value function implies the following policy function and resulting optimal capital function:

$$c(k) = \theta k^\alpha (1 - \beta\alpha), \qquad (7.10)$$

$$\tilde{k}(k) = \theta k^\alpha (\beta\alpha). \qquad (7.11)$$

The policy function describes optimal consumption c based on any given cap-ital stock k, while the analogue for capital, Equation 7.11, describes optimal $\tilde{k}$, given the current capital stock k.

7.1.1 COMPUTATION

In this very special case, we can solve the dynamic program analytically. We rarely have this luxury. Instead, we usually have to proceed numerically. Equa-tion 7.8 gives us a number of hints about how we can do this. We need to calculate $(\theta k^{\alpha} - \tilde{k})$ over a grid of different values of $\tilde{k}$; we then choose the 'best' outcome for $\tilde{k}$ (in the sense that it maximizes $\Gamma V(k)$); and finally, we do this for 'all' values of k.

Let's look at some code . . .

IterateVF.m

```matlab
function [TV optK] = IterateVF(V,maxK)
%-----------------------------------------------------
% PURPOSE: takes a potential value function V and
% performs an iteration, returning the updated
% proposed value function TV.  When TV=V, we
% have found the true value function. The scalar
% maxK represents the maximum possible amount of
% capital that can be consumed in one period
%-----------------------------------------------------
% INPUTS:    V : Nx1 vector of potential value
%                function
%         maxK : scalar of maximum capital
%                that can be consumed in a period
%-----------------------------------------------------
% OUTPUT:   TV : updated value function
%         optK : vector of optimal capital amounts
%-----------------------------------------------------

%----- (1) Basic Parameters --------------------
  Alpha   =   0.65; Beta   =   0.9; Theta   =   1.2;

  grid    =   length(V);
  K       =   linspace(1e-6,maxK,grid)';
  TV      =   zeros(length(V),1);

%----- (2) Loop through and create new value
%-----    function for each possible capital value
  for  k = 1:grid
     c                    = Theta*K(k)^Alpha-K(1:k);
     c(c<=0)              = 0;
     u                    = log(c);
     [TV(k) optK(k)]      = max(u + Beta*V(1:k));
  end
return
```

This code provides one iteration of the value function, after being passed a proposed value function V, and given an upper bound for capital (maxK). The first section of the code simply inputs our necessary parameters. The second section is the important part of this function. First, we loop over all possible values of k. This ensures that when we find our final solution, it will hold for all k. In this loop we calculate utility based on all possible values for $\tilde{k}$. In line 30, we maximize the current iteration of the value function; this returns TV(k) (the maximized value function) and optK(k) (the capital value associated with this maximum).

We can now use IterateVF to iterate! In the following code, we run through 10 value function iterations.

IterateGraph.m

```matlab
%-----(1) Set parameters, plot analytical
%          solution -----
Beta = 0.9; Alpha = 0.65; Theta = 1.2;
aB = Alpha*Beta;
K = linspace(1e-6,100,1000);

E = Alpha/(1-aB);
F = 1/(1-Beta)*(log(Theta*(1-aB)) ...
    + aB*log(aB*Theta)/(1-aB));
soln = E*log(K)+F;

plot(K,soln, '-k', 'LineWidth', 3)
axis([0 100 -15 10])
hold on

%----- (2) Plot 10 value function iterations ----
TV = [zeros(1000,1) NaN(1000,9)];
for iter = 1:10
    fprintf('Iteration number %d\n', iter)
    TV(:,iter+1)=IterateVF(TV(:,iter),100);
end

plot(K,TV)
xlabel('Amount of Capital', 'FontSize', 12)
ylabel('Value Function', 'FontSize', 12)
title('Value Function Iteration', 'FontSize', 14)
```

This script stores our 10 value function iterations in the matrix TV, along with the initial value function (which is just a vector of zeros).

Figure 7.1 shows these 10 value functions (the thin coloured lines), alongside the analytical solution (the thick black line). There is clearly more work to do—our final iteration is not that close to the target result.

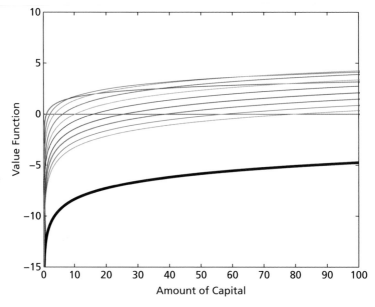

Figure 7.1 Convergence After 10 Iterations

Let's improve our script—by explicitly requiring the value function to converge. We will treat our function as having 'converged' when $V_{j+1}(k) \approx$
$V_j(k) \; \forall \; k$, where j just indicates the iteration number of the value function.[1]
To operationalize this in MATLAB, we use a while loop. We ask MATLAB to
keep iterating on the value function until the following convergence criterion
is met:

$$||V_{j+1}(k) - V_j(k)|| \leq \delta \; \forall \; k \qquad (7.12)$$

In the code that follows, δ is labelled `crit` (which we define as 0.01). At the
end of each iteration, we calculate $||V_{j+1}(k) - V_j(k)||$, which we call `conv`.

ConvergeVF.m

```
1  %----- Convergence to the Value Function --------
2  conv =   100;
3  crit =   1e-2;
4
5  K    =   linspace(1e-6,100,1000);
```

---

[1] We write converged in inverted commas here to imply that it is a slight abuse of nomenclature. In
numerical iterations we will never have true convergence of the value function. Rather, contiguous value
functions will move closer and closer to one another as we iterate towards infinity until the distance
between them is extremely small.

```
6   V       =   zeros(1000,1);
7   axis([0 100 -15 10])
8   hold on

9
10  cc      =   hot(70);
11  Iter =   0;

12
13  while conv>crit
14      Iter        =   Iter+1
15      [TV optK]   =   IterateVF(V,100);
16      conv        =   max(abs(TV-V))
17      plot(K,TV, 'color', cc(Iter,:))
18      V           =   TV;
19  end
```

The output from this code is presented in Figure 7.2; our value function does indeed converge.[2] What happens to the following if you start with an alternative $V_0(k)$? What if you use a more finely spaced capital grid?

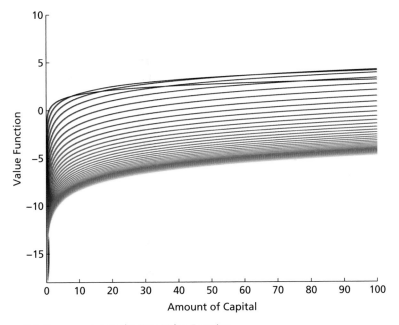

Figure 7.2 Convergence to the True Value Function

[2] In this graph we have used MATLAB's inbuilt 'colormap' hot, which results in higher iterations on the value function being 'hotter' colours.

7.1.2 THE POLICY FUNCTION

We have solved the value function—but what does this imply about optimal firm behaviour? More specifically, how much should the firm consume at a given point in time, and how much should it save? Fortunately, finding this *policy function*—that is c(k), or the mapping from capital to consumption—is reasonably straightforward.

You may remember that in the function IterateVF, we solved both for the value function at each point on our capital grid, as well as the corresponding optimal capital at this point. We return this optimal capital vector when we define the function IterateVF.m which returns the optimal capital vector as the output optK. Finally, when we run our optimal convergence code, we save two vectors for each iteration: TV, the value function, and optK, the optimal amount of capital.

Below we include code that uses this optimal capital vector to determine how future capital, $\tilde{k}$, should look given any specific value of k. While this is not the policy function *per se*, it is very close.[3] Figure 7.3b shows both the analytical (blue line) and numerically calculated (red line) optimal capital functions. This is exactly as defined in Equation 7.11. Exercise (i) asks you to generate the equivalent figures based on the *policy* function, Equation 7.10.

```
>> aB   = 0.65*0.9;   Theta = 1.2;   Alpha = 0.65;
>> plot(K,K(optK),K,aB*Theta*K.^Alpha, '--r',...
   'LineWidth', 3)
>> xlabel('Amount of Capital', 'FontSize', 12)
>> ylabel('Optimal k_{t+1}', 'FontSize', 12)
>> title('Policy Function: K Consumption',...
   'FontSize', 14)
```

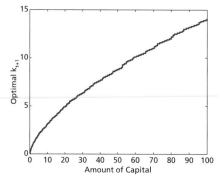

Figure 7.3a Calculated Best Path

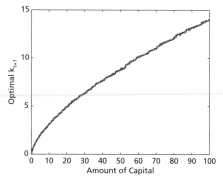

Figure 7.3b True Best Path

[3] Remember, the policy function is c(k). These figures, on the other hand, are $\tilde{k}(k)$.

7.2 **Policy Function Iteration**

In Section 7.1, we found numerical solutions to infinite horizon problems by iterating on the value function. However, this value function iteration is not necessarily computationally cheap. In some cases, this may not concern you—the 'traditional' value function iteration might work well for your particular problem. However, if you want to work with a very large state space, you may find that each additional iteration on the value function takes a long time to complete.

One very useful alternative to value function iteration is '*Howard's Improvement Algorithm*'—also known as policy function iteration. As we will see, this algorithm typically converges to the true policy function in many fewer steps than the value function iteration considered earlier.

As we saw with the code `ConvergeVF.m`, we required 66 iterations before V_{j+1} was close enough to V_j for us to consider that the function had 'converged'. What's more, in line 3 of this code, we defined 'convergence' as a situation where $||V_{j+1} - V_j|| < 0.01$. Were we to set a more rigorous convergence criterion, we would require (perhaps many) more iterations to solve the problem.[4]

In broad terms, a policy function iteration looks like the following, where the process is initialized by setting some initial arbitrary value function $V_j = V_0$, and defining some stopping criterion δ:

(i) Based upon V_j, determine optimal consumption for each k, giving a proposed 'policy function', $c_j(k)$;
(ii) Calculate the payoff associated with this policy function, $u(c_j(k))$;
(iii) Calculate the value of following this policy function forever, V_{j+1};
(iv) If $||V_{j+1} - V_j|| < \delta$, then stop, or else return to step (i) for another iteration.

The efficiency of this routine derives from step (iii)—in which we calculate the value function associated with following the policy function *forever*. In a traditional value function iteration, the calculated policy function is only followed for one period before again iterating to calculate V_{j+1}.

Policy function iteration generally takes many fewer steps than value function iteration, but there is one computationally heavy step: calculating the value function from the policy function. To see this, consider solving for the unknown V_j in the following:[5]

$$V_j = u(c_j(k)) + \beta Q_j V_j$$
$$\Rightarrow V_j = (\mathbf{I} - \beta \mathbf{Q}_j)^{-1} u(c_j(k)), \tag{7.13}$$

---

[4] For example, setting a convergence criterion of 1e-6 means that our problem now takes 153 iterations to converge.

[5] Here we borrow the notation of Judd (1998), and direct you to his discussion on pp. 411–17 should you be interested in further details.

where **I** is the identity matrix and **Q** is a matrix which keeps track of the capital stock associated with a given V_j.[6] At each step of Equation 7.13, which corresponds to item (iii) on the above list, V_j is calculated by matrix left division; this can be a computationally demanding process.

The function `IteratePolicy` implements the enumerated list above. The code makes use of MATLAB's `sparse` function, given that the matrix **Q** in Equation 7.13 is largely comprised of zeros; we discuss sparsity and the use of this function with more detail in Chapter 10.

IteratePolicy.m

```
1   function [V,opt] = IteratePolicy(V, maxK)
2   %--------------------------------------------------
3   % PURPOSE: takes an aribitrary value function
4   % V and iterates over the policy function c(k).
5   % At each step it calculates an updated policy
6   % function c_j(k), and a corresponding value
7   % function V_j(k), which is the value of.
8   % following c_j(k) forever.
9   %--------------------------------------------------
10  % INPUTS:    V : Nx1 vector of potential value
11  %                function
12  %         maxK : scalar of maximum capital
13  %                that can be consumed in a period
14  %--------------------------------------------------
15  % OUTPUT: V : updated value function at each
16  %             point
17  %       opt : optimal k for V
18  %--------------------------------------------------
19
20  %----- (1) Parameters --------------------------
21      Alpha   =   0.65; Beta   =   0.9;   Theta   =   1.2;
22      grid    =   length(V);
23      K       =   linspace(1e-6,maxK,grid)';
24      opt     =   NaN(grid,1);
25
26  %----- (2) Calculate optimal k for V -----------
27      for k   =   1:grid
28          c                   =   Theta*K(k)^Alpha-K(1:k);
29          c(c<=0)             =   0;
```

[6] In the case of a stochastic infinite horizon model, **Q** acts as a transition matrix describing the probability of each realization in the stochastic portion of the model. We do not go into that here, and refer you once again to resources such as Judd (1998) if you are interested in extending your code to include these types of details.

```
30      u                 =   log(c);
31      [V1,opt(k)]        =   max(u+Beta*V(1:k));
32    end
33
34    kopt  =   K(opt);
35    c     =   Theta*K.^Alpha-kopt;
36    u     =   log(c);
37
38  %----- (3) Invert k, u to find V_{j+1} ----------
39    Q     =   sparse(grid,grid);
40
41    for  k  =  1:grid
42       Q(k,opt(k))  =  1;
43    end
44    TV  =  (speye(grid)-Beta*Q)\u;
45    V   =  TV;
46  return
```

We can now loop over `IteratePolicy` until the numerical policy function
has converged. We provide a brief script below to do this. The first eight lines
set graphing parameters and graph the analytical solution, while the `for` loop
iterates over the policy function 7 times, starting with the defined $V_0 = 0$.
Figure 7.4 presents output, and it appears as if after only 7 iterations we are
already very close to the true policy function.

GraphPolicy.m

```
1  %----- (1) Setup parameters, pre-fill matrices ---
2  cmap    = cool(7);
3  V       = zeros(1000,1);
4  K       = linspace(1e-6,100,1000);
5  aB      = 0.65*0.9; Theta = 1.2; Alpha = 0.65;
6
7  %----- (2) Plot analytical solution to determine
8  %           performance ----
9  plot(K, aB*Theta*K.^Alpha, '-k','LineWidth',3)
10 hold all
11
12 %----- (3) Iterate using the Howard Algorithm,
13 %           plot each step ----
14 for l = 1:7
15     [V,k]      = IteratePolicy(V,100);
16     plot(K,K(k), 'color', cmap(l,:))
```

```
17  end
18
19  legend('Analytical', 'Iter 1', 'Iter 2', ...
20  'Iter 3', 'Iter 4', 'Iter 5', 'Iter 6', ...
21  'Iter 7', 'Location', 'NorthWest')
22  xlabel('Amount of Capital')
23  ylabel('Optimal k_{t+1}')
24  title('Policy Function Iteration & Consumption')
```

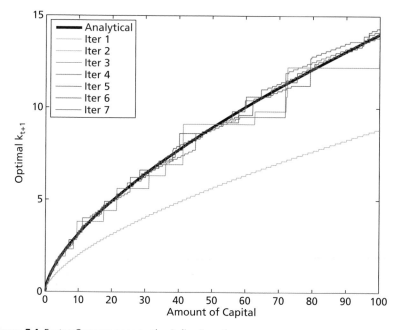

Figure 7.4 Faster Convergence to the Policy Function

7.3 **Estimating Infinite Horizon Models**

In Chapter 6, we spent considerable time discussing how to incorporate stochastic elements into our dynamic models, and, based on this, how to estimate structural parameters from data. We refrain from reproducing similar code in this chapter. However, we absolutely do not mean to suggest with this that simulation and estimation are not important processes in infinite horizon modelling. Quite the contrary! In any microeconometric application, estimation is a fundamental step, and certainly something that we as researchers will want to undertake. In a deeper sense, there is nothing really *different* between the two estimation processes (finite or infinite horizon), so we refer you back to Sections 6.4 and 6.5 for examples on how to do this.[7] We then refer you forward to Exercises (ii) and (iii) at the end of this chapter, which provides you some practice with writing your own simulation and estimation code.

7.4 **Review and Exercises**

Table 7.1 Chapter 7 Commands

Command	Brief description
linspace	A linearly spaced vector based on (user-defined) values
hot	A colourmap of black, red, and yellow for use in visual outputs
fsolve	Solves a system of equations
sparse	Store sparse matrix in a computationally more efficient way
speye	A sparse identity matrix, with only the diagonal stored in memory
legend	Add a legend to a plot

In this chapter we extend our dynamic analysis from Chapter 6, based on Bellman's Principle of Optimality, and value function iteration more generally, to an *infinite* horizon case. This means that we no longer have a final period from which to begin backwards iteration. However, as we have seen, this does not preclude us from solving these models. Reading this and the previous chapter and understanding and tweaking the code provided is a good start to writing and estimating these types of models. For those interested in learning more, we are happy to suggest a number of paths forward . . . .

(i) Much work on dynamic programming with an infinite horizon has been done in macroeconomics, for example considering country growth rates and investment decisions (starting with Ramsey). There is a large literature in this area, including resources on the application of these problems in computer languages (see for example the excellent references of Stachurski (2009) and Sargent and Stachurski (2013) in Python, or Collard (2013)'s online lecture notes on value function iteration in MATLAB with a macroeconomic focus).

---

[7] It is worth noting one important point about estimating infinite-horizon models: fortunately, we do not need an infinite number of periods of data. We can write our infinite-horizon model and, from this model, write moment conditions or a log-likelihood function. We can then use this to formulate an objective function, which we can use for estimation even on the finite data we have available.

(ii) So far in our models this chapter, we discretized our state space in such a way as to make it possible to tackle the problem in MATLAB. However, what if we face a problem whose state space is so large that it is not feasible to solve over a sufficiently fine grid? In this case, a very good solution is to solve the value function at only a reduced number of grid points, and *interpolate* in between these points. Typically, interpolation can be done using linear methods, least squares (where state variables are the explanatory variables), or splines. To see how this works with dynamic models, Keane and Wolpin (1994), who simulate and interpolate, is the place to start. More recent references include Rust (2000), who proposes the so called 'Parametric Policy Iteration'.

(iii) The points from Section 6.6 also hold for this chapter. Further numerical tools, ways to effectively grid and interpolate state space, and more complex stochastic elements may help in building and estimating more realistic and efficient models. We refer you to the references contained in Section 6.6.

EXERCISES

(i) In this chapter, we have used `ConvergeVF` to produce optimal capital graphs as Figures 7.3a and 7.3b. Write your own code which does the same thing, but generating optimal consumption graphs, in line with Equation 7.10.

(ii) In Chapter 6 we introduced a stochastic shock into the finite horizon model. Based on this example, write code to simulate the following model:

$$\max_{\{c_t\}_{t=1}^{\infty}} \sum_{t=1}^{\infty} \beta^{t-1} u(c_t) \qquad \text{subject to} \qquad k_{t+1} = \theta k_t^{\alpha} - c_t + \varepsilon_{t+1}.$$

(a) Assume that ε is structured as per the example in the previous chapter (a 50% chance of $+2$ and a 50% chance of -2). Simulate behaviour (capital and consumption) for 1,000 people over 10 time periods where the initial capital of each person is 100 units. Graph total consumption and remaining capital over time.

(b) Now, instead of assuming that each person starts with 100 units of capital, assume that each person starts with a uniform draw of between 50 and 100 capital units. How does this change optimal behaviour? Do you need to re-estimate the value function from part (a)?

(iii) Finally, imagine that the resulting capital and consumption paths from Exercise (ii) were provided to you as a dataset. Estimate the parameters θ and α (setting $\beta = 0.9$) from this data, assuming that the data generating process is that described in (ii). (*A useful hint*: The GMM code from the previous chapter—and the moments therein—are likely to be useful here.)

Appendix: Analytically Iterating the Value Function

In what follows, we use V_j to signify the value function, where the subscript $j \in [0, \infty)$ represents the iteration on the value function. Importantly, this number does not have any link to time periods, simply telling us how many times we have iterated over V, and hence how close we are to our solution. From Stokey and Lucas (1989), we know that under a relatively innocuous set of assumptions the contraction mapping theorem implies that as $j \to \infty$, $\Gamma V \to V$ (that is our value function will converge). To start the iterations, we define an initial value function:[8]

$$V_0(k) = 0 \ \forall \ k.$$

We treat V_0 as a proposed solution, where a proposed solution is only verified as the true solution if it is determined that $V_{j+1} = V_j$; otherwise, V_{j+1} becomes the new proposed solution, and iteration continues. So, starting from V_0, the first iteration is defined by maximizing the functional equation:

$$V_1(k) = \max_{\tilde{k}}\{\ln(c) + \beta V_0(\tilde{k})\} \qquad \text{s.t.} \qquad c = \theta k^\alpha - \tilde{k}. \qquad (7.14)$$

Here we use k and $\tilde{k}$ to denote capital in the current and subsequent periods respectively. In this case given that for all k the value of V_0 will be 0, it is optimal to consume all capital, giving a utility maximizing consumption of $c^* = \theta k^\alpha$. Substituting this optimal solution into our value function (Equation 7.14) gives:

$$V_1(k) = \ln(c^*) + \beta V(\tilde{k}^*)$$
$$= \ln(\theta k^\alpha) + \beta 0$$
$$= \ln \theta + \alpha \ln k, \qquad (7.15)$$

and, given that $V_1(k) \neq V_0(k)$, we know that our proposed V_0 is not the solution to the Bellman equation.

Now, having the result from the first iteration, we are able to iterate again, and continue the process of iteration until $V_j = V_{j+1}$, in which case we have arrived at our solution. For our second iteration, we continue as above:

$$V_2(k) = \max_{\tilde{k}}\{\ln(c) + \beta V_1(\tilde{k})\} \qquad \text{s.t.} \qquad c = \theta k^\alpha - \tilde{k}. \qquad (7.16)$$

Maximizing Equation 7.16 gives us a first order condition of the following form:

$$\frac{1}{\theta k^\alpha - \tilde{k}} = \frac{\beta \alpha}{\tilde{k}},$$

---

[8] This is arbitrary in the sense that starting the iteration from any resolvable value function will still lead us to the true solution.

which, by rearranging, implies that $\tilde{k}^* = \frac{\beta\alpha}{1+\beta\alpha}\theta k^\alpha$, and substituting into the flow equation that $c = \theta k^\alpha - \tilde{k}$ gives $c^* = \left(\frac{1}{1+\beta\alpha}\right)\theta k^\alpha$. Substituting these optimal values back into our value function gives that:

$$V_2(k) = \ln c^* + \beta V_1(\tilde{k}^*)$$

$$= \ln\left[\left(\frac{1}{1+\beta\alpha}\right)\theta k^\alpha\right] + \beta\left[\ln\theta + \alpha\ln\left(\frac{\beta\alpha}{1+\beta\alpha}\theta k^\alpha\right)\right]$$

$$= \alpha(1+\beta\alpha)\ln k + \ln\left(\frac{\theta}{1+\beta\alpha}\right) + \beta\ln\theta + \beta\alpha\left(\frac{\beta\alpha}{1+\beta\alpha}\theta\right)$$

$$= E_1\ln k + F_1$$

where in the second line the functional form for V_1 comes from Equation 7.15. E_1 and F_1 just denote constants and, once again, we can verify that $V_1(k) \neq V_2(k)$.

Similarly, we can iterate again to calculate $V_3(k)$:

$$V_3(k) = \max_{\tilde{k}}\{\ln(c) + \beta V_2(\tilde{k})\} \quad \text{s.t.} \quad c = \theta k^\alpha - \tilde{k},$$

$$= \max_{\tilde{k}}\{\ln(\theta k^\alpha - \tilde{k}) + \beta[\alpha(1+\beta\alpha)\ln\tilde{k} + F_1]\}$$

and here the relevant first order condition for the above equations is:

$$\frac{1}{\theta k^\alpha - \tilde{k}} = \frac{\beta\alpha(1+\beta\alpha)}{\tilde{k}}.$$

Rearranging this gives $\tilde{k} = \frac{\beta\alpha+\beta^2\alpha^2}{1+\beta\alpha+\beta^2\alpha^2}\theta k^\alpha$ and $c = \theta k^\alpha - \tilde{k} = \left(\frac{1}{1+\beta\alpha+\beta^2\alpha^2}\right)\theta k^\alpha$. We can then substitute these into our value function, giving:

$$V_3(k) = \ln c^* + \beta V_2(\tilde{k}^*)$$

$$= \ln\left[\left(\frac{1}{1+\beta\alpha+\beta^2\alpha^2}\right)\theta k^\alpha\right]$$

$$+ \beta\left[\alpha(1+\beta\alpha)\ln\frac{\beta\alpha+\beta^2\alpha^2}{1+\beta\alpha+\beta^2\alpha^2}\theta k^\alpha + F_2\right]$$

$$= \alpha(1+\beta\alpha+\beta^2\alpha^2)\ln k + F_2$$

$$= E_2\ln k + F_2$$

where once again E_2 and F_2 denote constants.[9]

---

[9] If you wish to do the algebra for F_2, feel free! If your algebra is correct (and we have not made any mistakes) you will find something like: $F_2 = \ln\left(\frac{\theta}{1+\beta\alpha+\beta^2\alpha^2}\right) + \beta\ln\left(\frac{\theta}{1+\beta\alpha}\right) + \beta^2\ln\theta + \beta^2\alpha\left(\frac{\beta\alpha}{1+\beta\alpha}\theta\right) + \beta\alpha\ln\left(\frac{\beta\alpha+\beta^2\alpha^2}{1+\beta\alpha+\beta^2\alpha^2}\theta\right)$.

Here, yet again, we see that $V_3(k) \neq V_2(k)$—however, we do start to see a pattern emerging. Indeed, were we to keep iterating *ad infinitum*, we would find that for each iteration j, the solution would look like $V_j(k) = E_j \ln k + F_j$. To resolve this fully, we could keep iterating, forming $V_4(k), V_5(k), \ldots$ In the limit, we can take advantage of the algebra of geometric series. For the first constant E_j, the limit is as follows:

$$\lim_{j \to \infty} E_j = \alpha[1 + \beta\alpha(1 + \beta\alpha + \beta^2\alpha^2 + \ldots + \beta^{j-1}\alpha^{j-1}] = \frac{\alpha}{1 - \alpha\beta}, \quad (7.17)$$

while for F we can break this down into a number of steps. From F_1 and F_2 we begin to see that the general form of F_j is:

$$F_j = \ln\left(\frac{\theta}{1 + \beta\alpha + \ldots + \beta^{j-1}\alpha^{j-1}}\right) + \beta \ln\left(\frac{\theta}{1 + \beta\alpha + \ldots + \beta^{j-2}\alpha^{j-2}}\right)$$

$$+ \ldots + \beta^{j-1}\ln\theta + \beta\alpha(1 + \beta\alpha + \ldots + \beta^{j-2}\alpha^{j-2}) +$$

$$\ln\left(\frac{\beta\alpha + \ldots \beta^{j-2}\alpha^{j-2}}{1 + \beta\alpha + \ldots \beta^{j-2}\alpha^{j-2}}\alpha\beta\theta\right) + \beta(\beta\alpha)(1 + \beta\alpha + \ldots + \beta^{j-3}\alpha^{j-3}) +$$

$$\ln\left(\frac{\beta\alpha + \ldots \beta^{j-3}\alpha^{j-3}}{1 + \beta\alpha + \ldots \beta^{j-2}\alpha^{j-2}}\alpha\beta\theta\right) + \ldots + \beta^{j-2}(\beta\alpha)\ln\left(\frac{1}{1 + \beta\alpha}\alpha\beta\theta\right).$$

This can be simplified into the sum of two geometric series. The first part of the above equation simplifies as:

$$\lim_{j \to \infty} \sum_{t=0}^{j-1} \beta^t \ln\left(\frac{1}{1 + \beta\alpha + \ldots + \beta^{j-1}\alpha^{j-1}}\theta\right) = \lim_{j \to \infty} \sum_{t=0}^{j-1} \beta^t \ln[\theta(1 - \beta\alpha)]$$

$$= \frac{1}{1 - \beta}\ln[\theta(1 - \beta\alpha)]$$

while the latter part of the equation simplifies to:

$$\lim_{j \to \infty} \sum_{t=0}^{j-2} \beta^t \beta\alpha(1 + \beta\alpha + \ldots + \beta^{j-2}\alpha^{j-2})\ln\left(\frac{\beta\alpha + \ldots \beta^{j-2}\alpha^{j-2}}{1 + \beta\alpha + \ldots \beta^{j-2}\alpha^{j-2}}\alpha\beta\theta\right)$$

$$= \lim_{j \to \infty} \sum_{t=0}^{j-2} \beta^t \frac{\beta\alpha}{1 - \beta\alpha}\ln(\beta\alpha\theta)$$

$$= \frac{1}{1 - \beta}\left[\frac{\beta\alpha}{1 - \beta\alpha}\ln(\beta\alpha\theta)\right],$$

in which case we have that

$$\lim_{j \to \infty} F_j = \frac{\ln[\theta(1 - \alpha\beta)]}{1 - \beta} + \frac{\beta\alpha\ln(\beta\alpha\theta)}{(1 - \beta\alpha)(1 - \beta)}.$$

Of course, this has been an awful lot of algebra, and we might be concerned that we have not actually found the closed form solution to this value function. Thankfully we have already come across a way we can check this solution: all we need to do is show that iterating once again on the above value function gives us an identical value function (a fixed point). Let's give it a try[10]...

$$V_{\infty+1}(k) = \max_{\tilde{k}}\{\ln(c) + \beta V_\infty(\tilde{k})\} \qquad \text{s.t.} \qquad c = \theta k^\alpha - \tilde{k}. \qquad (7.18)$$

As we have done above, we can form the first order condition for Equation 7.18, which gives:

$$\frac{1}{\theta k^\alpha - \tilde{k}} = \frac{\beta\alpha}{(1 - \beta\alpha)\tilde{k}}.$$

From here we can rearrange for $\tilde{k}^* = \beta\alpha\theta k^\alpha$, and $c^* = \theta k^\alpha(1 - \beta\alpha)$. If we substitute these optimal values into the value function we have:

$$V_{\infty+1}(k) = \ln c^* + \beta V_\infty(\tilde{k}^*)$$

$$= \ln\left[\theta k^\alpha(1 - \beta\alpha)\right] + \beta\left[\frac{\alpha}{1 - \alpha\beta}\ln(\beta\alpha\theta k^\alpha)\right.$$

$$\left. + \frac{\ln[\theta(1 - \alpha\beta)]}{1 - \beta} + \frac{\beta\alpha\ln(\beta\alpha\theta)}{1 - \beta}\right]$$

$$= \frac{\alpha}{1 - \alpha\beta}\ln k + \frac{\ln[\theta(1 - \alpha\beta)]}{1 - \beta} + \frac{\beta\alpha\ln(\beta\alpha\theta)}{(1 - \beta\alpha)(1 - \beta)}$$

and indeed, we find that $V_{\infty+1}(k) = V_\infty(k)$, indicating that we have iterated onto the true solution.

---

[10] And please excuse our abuse of notation in Equation 7.18.

Part IV

Nonparametric Methods

8   Nonparametric Regression

There are no straight lines in Nature.

Gaudí*

There are many straight lines in econometrics. Standard empirical techniques rely on strong assumptions about the economic problem under study. These assumptions simplify estimation and inference, but are rarely justified by economic theory or by other *a priori* considerations. Conclusions based on convenient but incorrect functional form assumptions can be misleading— estimated parametric models are often unable to capture the complex and highly non-linear patterns that we observe in microdata.

As the size of available datasets has grown, and as computing power has increased, so-called *nonparametric* methods have become increasingly popular. These methods make minimal assumptions about the process generating the data. They allow you to fit regression curves in a very flexible way—and, therefore, remain agnostic about the relationship between the dependent and explanatory variables.

There are a number of different methods available for nonparametric regression estimation—many more than we can cover here. In this chapter, we will introduce you to nonparametric smoothing regression in MATLAB. To do so, we will build, and fine-tune, our own Nadaraya-Watson kernel estimator. Kernel methods are simple to implement, are very flexible, and fundamental to a wide range of other nonparametric techniques.[1] At the end of the chapter, and in the exercises, we will briefly discuss how you can extend the approach to implement local linear regression, and point you in the direction of resources that deal with series estimation techniques.

8.1 Parametric Versus Nonparametric Approaches

At the risk of stating the obvious, the goal of a regression analysis is to produce a reasonable approximation of the relationship between some explanatory vari-

---

* Albert, Stuart, *When: The Art of Perfect Timing*, Jossey Bass (2013).
[1] Härdle (1991) argues that 'all smoothing methods are in an asymptotic sense essentially equivalent to kernel smoothing'.

ables X and a response variable y, given the data $\{X_i, y_i\}_{i=1,\ldots,N}$. The relationship between y_i and X_i is often modelled as:

$$y_i = m(X_i) + \varepsilon_i, \tag{8.1}$$

with the common assumption that $\mathbb{E}(\varepsilon \mid X) = 0$, and where $m(X)$ is referred to as the 'response function'.

There are three broad approaches that can be taken to model $m(X)$, which differ according to the degree of structure that is imposed on the response function.

(i) *The parametric approach*: We assume omniscience! Well, at least with regard to functional form. When taking a parametric approach, researchers assume that $m(X)$ takes a pre-specified functional form and is fully described by a small set of parameters that are to be estimated. For example, the linear model that we all know and love is a prime example of a parametric model:

$$y_i = X_i\boldsymbol{\beta} + \varepsilon_i. \tag{8.2}$$

(ii) *The nonparametric approach*: We assume nothing! Well, we assume very little.[2] When taking a nonparametric approach, the structure of $m(X)$ is assumed to be unknown and we let the data determine its form:

$$y_i = m(X_i) + \varepsilon_i. \tag{8.3}$$

(iii) *The semiparametric approach*: Semiparametric models sit in between the two extremes. A mixed model is assumed, which imposes some structure upon $m(X)$ and lets the data determine other parts of the function. For example, in some cases, we might be happy to model certain parts of the response function linearly. We could then partition the set of explanatory variables as $X = (W, Z)$, and assume that the regression relationship is linear in W and potentially non-linear in Z. This would lead to the specification that is called (imaginatively) the partially linear model:

$$y_i = W_i\boldsymbol{\beta} + m_z(Z_i) + \varepsilon_i. \tag{8.4}$$

We will tackle this approach in Chapter 9.

Different approaches will be suitable in different contexts, and there are many circumstances in which a fully parametric model seems to work well, or indeed is necessary—for example, the applications discussed throughout most of this book. However, when little is assumed about the underlying shape of the regression curve, a more flexible approach is needed. It is in these circumstances that the techniques discussed in this chapter become particularly useful.

---

[2] Assumptions are still made about the separability of, and distribution of, ε.

To illustrate the need for nonparametric estimation approaches, let's put MATLAB to use. Simulate a dataset $\{y_i, X_i\}_{i=1,...,N}$, where the data generating process producing y_i takes the form:

$$y_i = \sin(X_i) + \varepsilon_i, \tag{8.5}$$

with $\varepsilon \sim N(0, 1)$.

This is done at the command line as follows:

```
>> X = [0.1:0.1:10]';
>> y = sin(X) + randn(length(X),1);
```

The first line of code creates a 100-by-1 vector of the explanatory variable X. The second line of code generates the dependent variable y as the sine function of X plus a random normal error term.

Although OLS provides the best linear approximation to $m(\mathbf{X})$, it is way off the true response function in this example. Using basic matrix operations to calculate the OLS coefficients (see Chapter 1) and plotting the output shows this clearly.

```
>> X_ = [ones(length(X), 1) X];
>> betaOLS = (X_'*X_)\(X_'*y)


betaOLS =

    0.7572
   -0.0897


>> figure;
>> plot(X, X_*betaOLS, X, sin(X));
>> legend('OLS', 'm(X)')
>> hold on
>> scatter(X,y)
```

This recovers the OLS estimate, betaOLS, and then plots the fitted and true response functions. A scatter plot of the simulated data is added to the figure. Our OLS model is hopelessly misspecified (Figure 8.1). Surely we can do better than this?!

In this chapter, we will use nonparametric regression methods to achieve a more accurate approximation of the underlying regression function. However, it is worth noting that there are trade-offs involved when opting for non-parametric methods. Parametric estimation is a *finite-dimensional* problem—the aim is to recover the finite set of parameters that index the (assumed) parametric functional form. Nonparametric estimation problems, on the other hand, are *infinite-dimensional*—no restrictions are placed on the structure of $m(X)$ and so we could require an infinite number of parameters to specify the function correctly. This can create problems for the interpretation of regression

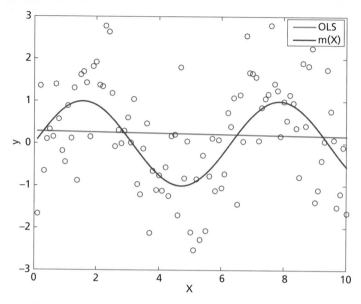

Figure 8.1 OLS Misspecification

output (it can be pretty difficult to get your head around an infinite dimensional object!) and also for the convergence of estimates to the truth. Parametric estimates typically converge at rate $\sqrt{N}$ but nonparametric estimates will converge at a slower rate than this, especially when the dimension of $\mathbf{X}$ is large.[3]

Nonparametric methods will not provide the right approach in every circumstance. However, they are very useful in a wide variety of empirical contexts. They should, therefore, be a part of every applied researcher's toolkit.

8.2 **Kernel Regression**

In this chapter, we focus on nonparametric univariate regression. The challenges that arise in the context of multivariate kernel regression are addressed in Chapter 9.

Imagine that we want to estimate the response function at a particular point x_0, $m(x_0)$. One basic estimator of $m(x_0)$ is simply the average of the y_i for respondents with X_i close to x_0. This is the central intuition behind kernel regression. Under the kernel approach, the estimator of $m(x_0)$ is constructed

---

[3] The famous statisticians Ronald Fisher and Karl Pearson had many a disagreement over (among many other things) which approach, parametric or nonparametric, was superior. Fisher objected to nonparametric methods because of their relative lack of efficiency, whilst Pearson was much more concerned with questions of specification.

as a weighted average of the response variable, y. Formally, the estimate $\hat{m}(x_0)$ is constructed as:

$$\hat{m}(x_0) = \frac{\sum_{i=1}^{N} W_i(x_0) y_i}{\sum_{i=1}^{N} W_i(x_0)} \tag{8.6}$$

where $\{W_i(x_0)\}_{i=1,\ldots,N}$ is the sequence of weights.

Most specifications for W_i embody the idea that one wants to give *less* weight to respondents whose X_i are *further* from x_0. This is where the kernel function, $K(\cdot)$, enters the story. A kernel function attaches greatest weight to observations that are close to x_0 and then gradually less weight to observations that are further away. All kernel functions are continuous, bounded, and symmetric real functions that integrate to one, and they imply weights of the form:

$$W_i(x_0) = K\left(\frac{X_i - x_0}{h}\right) = K_h(X_i - x_0) \tag{8.7}$$

where the scaling factor, h, is referred to as the 'bandwidth'.

The use of a kernel weighting function leads to the Nadaraya-Watson (NW) estimator. This estimator was proposed by (suprise, suprise!) Nadaraya (1964) and Watson (1964), and takes the form:

$$\hat{m}_h(x_0) = \frac{\sum_{i=1}^{N} K_h(X_i - x_0) y_i}{\sum_{i=1}^{N} K_h(X_i - x_0)}. \tag{8.8}$$

A number of different kernel functions are used in applied work. Table 8.1 gives the most common kernel functions, their formulae, and the integral of its square, b_K (which is needed to calculate the pointwise standard error— see Exercise (i)). Researchers must also choose the value of the bandwidth, with the only restriction being that $h > 0$. The choice of bandwidth is very important.

Let's revisit the regression problem of Section 8.1. Here we will construct an estimator using the Gaussian kernel function and a sensible guess for the bandwidth. (We will tackle methods for optimally choosing K and h in due course.)

Table 8.1 Common Kernel Functions

Kernel function	Mathematical formula	b_K
Epanechnikov	$K_h(u) = \frac{3}{4\sqrt{5}}(1 - \frac{1}{5}u^2)\mathbf{1}(|u| \leq 1)$	3/5
Gaussian	$K_h(u) = \frac{1}{\sqrt{2\pi}} e^{-\frac{1}{2}u^2}$	$1/(2\pi^{1/2})$
Quartic	$K_h(u) = \frac{15}{16}(1 - u^2)^2 \mathbf{1}(|u| \leq 1)$	5/7
Triangular	$K_h(u) = (1 - |u|)\mathbf{1}(|u| \leq 1)$	2/3
Uniform	$K_h(u) = \frac{1}{2}\mathbf{1}(|u| \leq 1)$	1/2

First, we must construct the kernel function. We could (as per usual) define the kernel function in a separate program file. However, here we will use MATLAB's *anonymous function* capabilities to create $K_h(u)$ without having to define it outside of the main body of code. Just like a 'normal' function, anonymous functions accept inputs and return outputs. However, anonymous functions can only contain a single executable statement. The anonymous function for the Gaussian kernel is:

```
>> GaussKernel = @(u) exp(-(u.*u)/2)/sqrt(2*pi);
```

The variable `GaussKernel` is the *function handle*, the callable association to the kernel function. The @ operator creates the function handle. The function inputs, here just **u**, are enclosed by the parentheses. The Gaussian kernel function accepts the vector **u** and returns a vector of the normal density values at each element of **u**. In the regression exercise, $\mathbf{u} = (\mathbf{X} - x_0)/h$.

Now, to the bandwidth! Bowman and Azzalini (1997) and Silverman (1986), among many others, provide 'plug-in' choices (that is, sensible guesses) for the bandwidth. Here we will use the formula given by Bowman and Azzalini (1997), which can be implemented in MATLAB as:

```
>> N = length(X);
>> hx=median(abs(X-median(X)))/0.6745*(4/3/N)^0.2;
>> hy=median(abs(y-median(y)))/0.6745*(4/3/N)^0.2;
>> h=sqrt(hy*hx);
```

Armed with the Gaussian kernel function and a choice for the bandwidth, we can now use the NW estimator to recover the response function that generated the data we used at the beginning of the chapter.

Kernel regression proceeds pointwise. You must choose each value of x at which you want to evaluate the regression function. The snippet of code below evaluates $\hat{m}_h(x)$ at $\{X_i\}_{i=1,\dots,N}$, the values of x in the dataset. For each X_i, u (on line 3) records its distance from every other observation in the dataset, scaled by h. Line 4 uses the anonymous function `GaussKernel` (created above) to evaluate the kernel weights associated with each observation. Finally, we use the NW estimator to bring together a weighted sum of the observed y to form the nonparametric estimate, $\hat{m}_h(X_i)$.

```
1  yhat = NaN(N,1);
2  for i = 1:N
3      u = (X - X(i))/h;
4      Ku = GaussKernel(u);
5      yhat(i) = sum(Ku.*y)/sum(Ku);
6  end
```

We can compare the output of the OLS and NW regressions by plotting the true response function ($\sin(x)$), our nonparametric estimate (`yhat`), and the

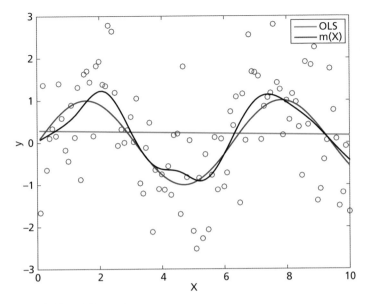

Figure 8.2 Basic Kernel Regression

OLS predicted values. This is shown in Figure 8.2. It is clear that, although there are still some wiggles, the nonparametric estimate provides a much better approximation of the true response curve.

```
figure;
plot(X, sin(X), X, yhat, X, X_*betaOLS);
legend('m(X)', 'NW estimator', 'OLS');
```

There remains some discrepancy between the true curve and the nonparametric estimate. Some of this inaccuracy derives from our arbitrary choices for the bandwidth and the kernel function. We have not yet optimized these choices, but we will do so shortly.

8.2.1 CELL ARRAYS AND STRUCTURE ARRAYS

In section 8.2.2, we will need to store the regression output associated with different kernels and bandwidths. That is, we will need to store a family of related objects. MATLAB provides two convenient ways to handle this: structure arrays and cell arrays. These are alternative data structures that allow you to store multiple data types, and data of different sizes, in the same variable. Let's take a quick detour to introduce these now.

Let's first introduce *structure arrays*. To see how structures work, let's store the nonparametric regression output from the exercise earlier in this chapter. Of course, we can store the output, kernel, and bandwidth used as different variables. However, it may be useful to store these together. Like a paper clip holding together related documents, a structure holds together different data pieces that relate to each other.

The following code creates a structure called `Kresults`, which stores our kernel output. As you can see, the syntax is very simple.

```
>> Kresults.kernel = 'Gaussian';
>> Kresults.h = h;
>> Kresults.yhat = yhat;
```

To examine a structure's contents, just type its name at the command line.

```
>> Kresults

Kresults =

    kernel: 'Gaussian'
         h: 1
      yhat: [100x1 double]
```

The structure `Kresults` has three components: `kernel`, `h`, and `yhat`. These are called 'fields'. We can list all the field names of a structure using the function `fieldnames`:

```
>> fieldnames(Kresults)

ans =

    'kernel'
    'h'
    'yhat'
```

You can add new fields to the structure at any time. For example, let's add a new field to record the date that the regression was performed:

```
>> Kresults.date=date;
>> Kresults

Kresults =

    kernel: 'Gaussian'
         h: 1
      yhat: [100x1 double]
```

```
    date: '03-Jul-2014'

>> fieldnames(Kresults)

ans =

    'kernel'
    'h'
    'yhat'
    'date'
```

This adds an additional field `Kresults.date` in dd-mm-yyyy format. Inspecting the structure and its field names shows that this has resulted in the additional date field being added.

The elements of a structure can be operated on just like any other MATLAB object. For example:

```
>> Kresults.h*10

ans =

    10
```

An array of structures can be created to represent multiple objects. Let's create an array of structures to hold the estimates associated with different choices for the bandwidth and kernel function. For example, imagine that we are interested in also storing the results associated with a bandwidth of 6 and a uniform kernel function:

```
>> Kresults(2).kernel = 'Uniform';
>> Kresults(2).h = 6;
>> Kresults

Kresults =

1x2 struct array with fields:
    kernel
    h
    yhat
    date
```

All structures in the array have the same number of fields, with the same field names. Any unspecified fields in new structures within the array simply contain empty arrays. For example, the fields `yhat` and `date` are left empty for the second record until they are specified (as we shall do later in the chapter).

```
>> Kresults(2).yhat

ans =

    []
```

Cell arrays also allow us to store together data of different types and sizes. To see how they work, imagine that you want to store the names of the various estimation approaches: parametric, nonparametric, semiparametric. You might think that you could create an array of strings in MATLAB as follows:

```
>> approaches = ['parametric', 'nonparametric',...
                 'semiparametric'];
```

Sadly, this does not work.

```
>> disp(approaches)
parametricnonparametricsemiparametric
```

Oh dear—this is not what we were expecting! Why did this happen? In MAT-LAB, a string is stored as an array of characters. Therefore, if you try to create an array of strings as above, MATLAB simply concatenates the separate character arrays together to form one character array—that is, one long string. MATLAB's cell arrays allow us to avoid this problem. They allow you to store an array of *character* arrays—that is, they let you store an array of strings!

The main difference between creating a normal array and creating a cell array is the use of brackets—you switch from square to curly brackets to make a cell array. Not so difficult, really!

```
>> approaches = {'parametric', 'nonparametric',...
                 'semiparametric'}

approaches =
    'parametric'    'nonparametric'    'semiparametric'
```

There is a also a difference in how data is accessed from a cell array. To see this, try changing the second string in our cell array to 'excellent!' You might think that you would do so as follows...

```
>> approaches(2) = 'excellent';
??? Conversion to cell from char is not possible.
```

When you index a cell array with round parentheses, you are referring to sets of cells *not* to the contents within those cells. You can think of a cell array as a series of boxes (the cells). To pull out a specific box, you use round brackets. To access what is inside the box, you need to use curly brackets. You can see this by examining the data classes of the objects `approaches(2)` and `approaches{2}`.

```
>> class(approaches(2))

ans =

cell

>> class(approaches{2})

ans =

char
```

To access and change the *contents* of a cell, you need to index using curly brackets.

```
>> approaches{2} = 'excellent!'

approaches =
    'parametric'    'excellent!'    'semiparametric'
```

In the section, we use cell arrays to change the kernel function that we use for our nonparametric regressions.

8.2.2 KERNEL AND BANDWIDTH CHOICE

Let's now record the kernel estimates as the bandwidth and choice of kernel function are varied. This will highlight the impact of these choices on the quality of the regression output.

Let's consider three choices for the kernel function: the Gaussian kernel, the Epanechnikov kernel, and the Uniform kernel. The MATLAB anonymous function equivalents of these kernels are given in Table 8.2.

Table 8.2 Kernel Functions in MATLAB

Kernel function	Matlab code
Gaussian	GaussKernel = @(u) exp(-(u.*u)/2)/sqrt(2*pi)
Epanechnikov	EpanKernel = @(u) 0.75*(1 - (u.*u)).*(abs(u)<=1)
Uniform	UniKernel = @(u) 0.5.*(abs(u) <= 1)

The differences between the kernel weighting functions are easily seen once plotted over some range (Figure 8.3). This is done at the command line with the following code.

```
>> u = [-5:0.01:5]';
>> plot(u, GaussKernel(u), u, EpanKernel(u), u,...
   UniKernel(u));
>> legend('Gaussian', 'Epanechnikov', 'Uniform');
>> xlabel('u');
>> ylabel('K(u)');
```

The uniform kernel function is discontinuous, which might not be desirable if you believe that the underlying response function of interest is continuous—it is necessary for the kernel function to be continuous to obtain a continuous estimate. The Gaussian and Epanechnikov kernel functions are both popular choices for applied work—both are continuous, although the Gaussian function places positive weight on the full support of X and the Epanechnikov kernel does not have a derivative where $|u| = 1$, which can be a disadvantage. (The Epanechnikov does have some further desirable efficiency properties that are discussed further below.)

For each of these three kernel functions, we will consider four potential choices for the bandwidth, $h = [0.001; 3; 6; 40]$.

The function Kreg takes in the data, the points at which we want to evaluate our kernel regression and, if we so desire, choices for the bandwidth and kernel function. It then outputs the estimated regression values, yhat, and the bandwidth used, h.

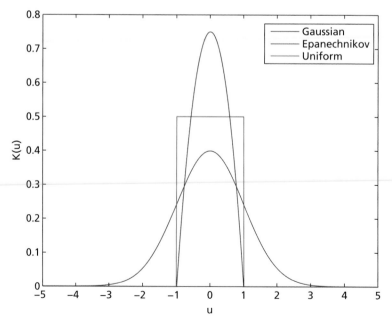

Figure 8.3 Alternative Kernel Weighting Functions

Kreg.m

```
1   function [yhat,h] = Kreg(X,y,x,h0,func)
2   %-------------------------------------------------
3   % PURPOSE: performs the kernel regression of y on
4   %             (univar) X
5   %-------------------------------------------------
6   % USAGE: yhat = Kreg(X,y,x,h0,func)
7   % where: y : n-by-1 dependent variable
8   %          X : n-by-1 independent variable
9   %          x : N-by-1 points of evaluation
10  %          h0 : scalar bandwidth
11  %          func : kernel function
12  %-------------------------------------------------
13  % OUTPUT: h : bandwidth used
14  %          yhat : regression evaluated at each
15  %                   element of x
16  %-------------------------------------------------
17  n = length(y);
18  N = length(x);
19
20  %--- (1) Set bandwidth if not supplied ----------
21  if nargin < 4
22    % suggested by Bowman and Azzalini (1997)
23    hx=median(abs(X-median(X)))/0.6745*(4/3/n)^0.2;
24    hy=median(abs(y-median(y)))/0.6745*(4/3/n)^0.2;
25    h=sqrt(hy*hx);
26  else
27    h = h0;
28  end
29
30  %--- (2) Set Kernel function if so desired ------
31  if nargin < 5
32      % Gaussian as default
33      krnl = @(u) exp(-(u.*u)/2)/sqrt(2*pi);
34  else
35    K = {'Gaussian', 'Uniform', 'Epanechnikov'};
36
37    if strcmpi(func, K{1})
38        % Gaussian
39        krnl = @(u) exp(-(u.*u)/2)/sqrt(2*pi);
40    elseif strcmpi(func, K{2})
41        % Uniform
42        krnl = @(u) 0.75*(1 - (u.*u)).*(abs(u) <= 1);
```

```
43    elseif strcmpi(func, K{3})
44        % Epanechnikov
45        krnl = @(u) 0.5.*(abs(u) <= 1);
46    else
47        % --> if there's been a mistake- Gaussian
48        krnl = @(u) exp(-(u.*u)/2)/sqrt(2*pi);
49    end
50    end
51
52    %--- (3) Perform kernel regression! ------------
53    yhat = NaN(N,1);
54    for i = 1:N
55        u = (X - x(i))/h;
56        Ku = krnl(u);
57        yhat(i) = sum(Ku.*y)/sum(Ku);
58    end
59
60    return
```

It is optional for the user to specify the bandwidth and kernel when using Kreg. This is made possible by the MATLAB function nargin. This is found on line 21 of Kreg. nargin returns the number of arguments passed in the call to the function. If the number of arguments passed to the function is fewer than four, then Kreg uses the default plug-in bandwidth that was used previously. If, however, nargin < 4 evaluates as false, Kreg uses h0.

A similar sequence of actions occurs with the choice of kernel function—if the number of arguments passed to the function is fewer than five, the Gaussian kernel is used. If not, MATLAB's string compare function, strcmpi, is used to compare the user-supplied string with the strings contained in the cell array K.[4] Keep an eye out for the use of a cell array in line 35.

Having specified the bandwidth and kernel, the final part of the function implements the NW estimator (which we walked through earlier in the chapter) at each of the points in the user-supplied vector x.

8.2.3 INFLUENCE OF KERNEL AND BANDWIDTH

The script below creates a structure, Kresults, to record the output associated with different combinations of the kernel function. It calls on the function Kreg to perform the regressions.

---

[4] strcmpi is case insensitive. You can use strcmp to compare strings with case sensitivity.

The code begins using the `struc` syntax to create an empty structure, `Kresults`, with three fields (name, h, and yhat). We populate this structure using a double loop to cycle through our three choices of kernel function and four choices for the bandwidth. The names of the kernel function and the value of the bandwidth are recorded, before the kernel regression is performed using `Kreg`.

KregStruc.m

```
1   % Script: Kernel Regression with Structures
2   clear; clc;
3
4   %----- (1) Simulate data ---------------------
5   X = [1:100]';
6   y = sin(0.1*X) + randn(length(X),1);
7
8   %----- (2) Prepare structure array -----------
9   K = {'Gaussian', 'Uniform', 'Epanechnikov'};
10  % K is a 'cell array'
11  K_ = length(K);
12  h = [1e-3; 3; 6; 40];
13  h_ = length(h);
14  Kresults = struct('name',[],'h',[],'yhat',[]);
15
16  %----- (3) Perform kernel regression! ----------
17  for k =1 : K_ % cycle over kernel choices
18
19      kernel = K{k};
20
21    for hh = 1 : h_ % cycle over bandwidth choices
22
23      bandwidth = h(hh);
24
25      % record name and bandwidth
26      Kresults((k-1)*h_ + hh).name = kernel;
27      Kresults((k-1)*h_ + hh).h = bandwidth;
28
29      % perform kernel regression
30      Kresults((k-1)*h_ + hh).yhat = ...
31              Kreg(X, y, X, bandwidth, kernel);
32
33    end
34  end
```

```
35
36    %----- (4) Plot output  ------------------------
37    % Bandwidth
38    figure;
39    for hh = 1 : h_
40        subplot(2, 2, hh);
41        scatter(X,y);
42        hold on
43        plot(X, Kresults(hh).yhat,'Color','r',...
44            'LineWidth', 1.5);
45        hold off
46    end
47
48    % Kernel function
49    figure;
50    for k = 1 : K_
51        subplot(3, 1, k);
52        scatter(X,y);
53        hold on
54        plot(X, Kresults((k-1)*h_+3).yhat,'Color','r');
55        hold off
56    end
```

Figures 8.4 and 8.5 show the output of the nonparametric regressions and a scatter plot of the data for the different choices of the kernel function (using a bandwidth of 6), and of the different choices of the bandwidth (with the Gaussian kernel).

It is clear that the choice of bandwidth has a huge influence on the regression output, while the choice of kernel function does not influence the general shape of the curve too significantly. In fact, in large samples, the choice of kernel has negligible impact, although it can be shown that the optimal kernel (in the sense of minimizing mean squared error (MSE)) is the Epanechnikov, although the advantage is minimal (Hodges and Lehmann (1956)).

8.3 Cross Validation

The choice of bandwidth has caused much head scratching. The bandwidth is clearly very important for the overall quality of our regression output, but it is hard to work out how exactly to choose its value optimally. There is a trade-off between setting h small to reduce bias (there would be no bias if we simply interpolated between data points) and setting h large to increase smoothness.

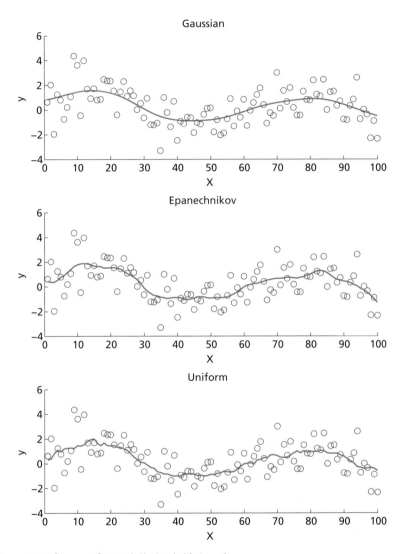

Figure 8.4 Influence of Kernel Choice (with $h = 6$)

This trade-off is clear mathematically in the derivation of the mean squared error of the kernel estimator. (See the references at the end of this chapter for a further discussion of this point.)

A popular and practical approach to bandwidth selection is *cross validation*. This method makes the choice of bandwidth directly dependent on the data. We want the choice of h to minimize the sum of squared errors:

$$\sum (y_i - \hat{y}_i)^2 = \sum \hat{\varepsilon}_i^2. \qquad (8.9)$$

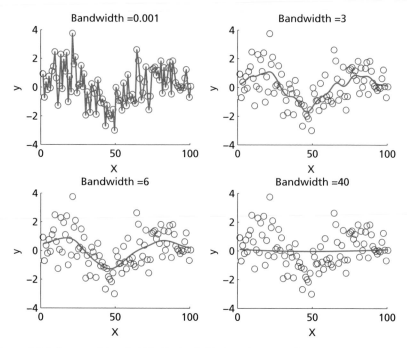

Figure 8.5 Influence of Bandwidth Choice (with Gaussian Kernel)

However, choosing h to minimize this sum is *not* a sensible thing to do! As h shrinks, the within-sample fit of the model improves and $\sum \hat{\varepsilon}_i^2$ decreases. In fact, as h approaches zero, $\hat{m}_h(X_i)$ collapses to y_i to obtain a perfect fit. So, simply picking h to minimize $\sum \hat{\varepsilon}_i^2$, would result in $h^* = 0$ and an interpolation of observed data points—which is certainly not what we are aiming for.

Rather than pick h to minimize the sum of squared residuals, we can choose h to minimize the sum of squared 'leave-one-out' residuals:

$$CV = \sum_{i=1}^{N} (y_i - \hat{m}_{\neg i}(X_i))^2, \tag{8.10}$$

where $\hat{m}_{\neg i}(X_i)$ is the kernel estimate at X_i obtained by omitting the ith observation.

Formally, we obtain the cross validation bandwidth by solving the following optimization problem:

$$\min_{h} \sum_{i=1}^{N} (y_i - \hat{m}_{h,\neg i}(X_i))^2 \tag{8.11}$$

subject to the constraint that $h \geq 0$.

From Equation 8.11, it looks like we might have to run N nonparametric regressions to select h. However, luckily, cross validation is not as computationally intensive as it might first appear. It can be shown that:

$$y_i - \hat{m}_{\neg i}(X_i) = \frac{y_i - \hat{m}(x_i)}{1 - \frac{K_h(X_i - X_i)}{\sum_i K_h(X_i - X_j)}}. \tag{8.12}$$

Therefore, the value of the objective function, for a given choice of h, only requires a single computation of the regression function. Phew!

We can easily perform cross validation in MATLAB with the help of, you guessed it, our favourite friend fmincon. This is, of course, just another constrained optimization problem. To proceed, we create first a function that returns the sum of squared errors associated with the leave-one-out kernel regression, given a particular value for the bandwidth h. We can then use fmincon to minimize the value of this function.

The function MSE accepts the data and a particular bandwidth as inputs and outputs the sum of squared 'leave-one-out residuals'. We begin by declaring the Gaussian kernel function (feel free to change this to your favourite one). In the for loop, MATLAB cycles over each of the N observations and estimates the regression function using the supplied bandwidth h. The kernel weight on the ith observation and the sum of the kernel weights at each point are saved so we can calculate the leave-one-out MSE as in Equation 8.12.

MSE.m

```matlab
1  function value = MSE(X,y,h)
2  %------------------------------------------
3  % PURPOSE: calculate the MSE for the leave-one-
4  %          out kernel reg
5  %------------------------------------------
6  % INPUTS:  X : Nx1 vector of independent variable
7  %          y : Nx1 vector of dependent variable
8  %          h : candidate bandwidth
9  %------------------------------------------
10 % OUTPUT: value : mean squared error of
11 %                 leave-one-out
12 %------------------------------------------
13
14 %----- (1) Declare kernel function -----------
15 krnl = @(u) exp(-(u.*u)/2)/sqrt(2*pi);
16
17 %----- (2) Summary vectors ------------------
18 N                  = length(X);
```

```
19   yhat = NaN(N,1);
20   weight_i = NaN(N,1);
21   weight_sum = NaN(N,1);
22
23   %-----(3) Perform kernel regression & save
24   %          weights ------
25   for i = 1:N
26       u = (X - X(i))/h;
27       Ku = krnl(u);
28       weight_i(i) = Ku(i);
29       weight_sum(i) = sum(Ku);
30       yhat(i) = sum(Ku.*y)/sum(Ku);
31   end
32
33   %----- (4) Calculate leave-1-out MSE -----------
34   residual = (y-yhat)./(1-(weight_i./weight_sum));
35   value = sum(residual.^2);
36
37   return
```

With the MSE function to hand, we have all the ingredients to calculate the optimal cross validation bandwidth. Given some starting value h_0, we pick the bandwidth to minimize the objective function MSE using fmincon subject to the constraint that h is weakly positive. This is done in the lines of code below. We supply fmincon with the plug in starting value, h0, and minimize the function MSE by varying h only. The placeholders for the equality and inequality constraints are simply made up of empty braces, although we place a lower bound of 0 on the choice of h.

```
>> N = length(y);
>> hx=median(abs(X-median(X)))/0.6745*(4/3/N)^0.2;
>> hy=median(abs(y-median(y)))/0.6745*(4/3/N)^0.2;
>> h0=sqrt(hy*hx);
>> options = optimset('Display', 'off', ...
               'Algorithm', 'sqp');
>> [opth,~,exitflag] = fmincon(@(h)MSE(X,y,h), h0,...
                       [],[],[],[], 0, [],[],...
                       options)

opth =
     6.5241
exitflag =
     1
```

8.3.1 TRIMMING

We have, so far, treated every observation symmetrically. However, there are times when it might be sensible to downweight the influence of certain observations.

To remind you, the NW estimator takes the form:

$$\hat{m}_h(x_0) = \frac{\sum_{i=1}^{N} K_h(X_i - x_0)y_i}{\sum_{i=1}^{N} K_h(X_i - x_0)}. \qquad (8.13)$$

The denominator of this expression is $\hat{f}(x_0)$—the kernel estimate of the density of the regressor at x_0. There might be some places, especially in the tails, where $\hat{f}(x_0)$ is very small, causing erratic estimates of $\hat{m}_h(x_0)$.

With trimming, one downweights or excludes observations at which $\hat{f}(X_i) < \delta$, with $\delta \to 0$ as $N \to \infty$. The cross validation criterion with trimming takes the form:

$$CV = \sum_{i=1}^{N}(y_i - \hat{m}_{\neg i}(X_i))^2 \cdot \pi(X_i), \qquad (8.14)$$

where $\pi(X_i)$ is the trimming function. For example, one might choose not to place any weight on observations outside the 5th and 95th percentiles, in which case $\pi(X_i) = 0$ for these observations and $\pi(X_i) = 1$ otherwise. We invite you to add a trimming function in Exercise (vi).

8.4 **Local Linear Regression**

The kernel estimator is a 'local constant' estimator: $m(x)$ is assumed to be constant in the local neighbourhood of x. This approach can be generalized to let $m(x)$ be linear in the close neighbourhood of x. We will cover this topic very briefly as, having mastered the standard kernel approach, implementing this estimator in MATLAB is straightforward.

The local linear estimator has better properties at the boundary than the standard NW estimator and has zero bias if the true regression is linear. However, before doing away with the NW estimator, it is important to note that the local linear estimator does not always dominate. When the regression function is rather flat, the standard kernel estimator tends to do better in simulations. Yet, when the underlying function is more curvy, the local linear might be a better bet.

To fix ideas, note that the standard kernel estimator can be expressed as:

$$\hat{m}(x_0) = \underset{m_0}{\text{argmin}} \, K_h(x_0 - X_i)(y_i - m_0)^2. \qquad (8.15)$$

Rather than just picking a constant, the local linear estimator fits a linear regression in the close neighbourhood of x_0:

$$\hat{m}^{LL}(x_0) = \underset{a_0,b_0}{\operatorname{argmin}} K_h(x_0 - X_i)(y_i - a_0 - b_0(X_i - x_0))^2. \qquad (8.16)$$

We use the regressor $(X_i - x_0)$, rather than just X_i, to ensure that $m(x) = E(y_i|X_i = x)$. Conveniently, once we have recovered the estimates $\hat{a}_0(x)$ and $\hat{b}_0(x)$, we can set $\hat{m}(x) = \hat{a}_0(x)$ and $\frac{\partial \hat{m}(x)}{\partial x} = \hat{b}_0(x)$.

Solving this optimization problem yields an explicit expression for the local linear estimator (see Fan and Gijbels, 1996):

$$\hat{\beta} = (\tilde{X}'W\tilde{X})^{-1}\tilde{X}'Wy \qquad (8.17)$$

where

$$\tilde{X} = \begin{bmatrix} 1 & x_0 - X_1 \\ \vdots & \vdots \\ 1 & x_0 - X_N \end{bmatrix} \qquad (8.18)$$

and $W_{ii} = K_h(x_0 - X_i)$. This expression makes clear the connection to Weighted Least Squares. It is simple really! Exercise (v) asks you to extend the local linear estimator to a local polynomial estimator of degree p.

8.4.1 IMPLEMENTING IN MATLAB

Now that you have the NW estimator under your belt, coding the LL estimator is easy. The basic machinery is the same as that of Kreg. The only parts of the function that need to change are the declaration of the function outputs and the implementation of the local linear estimator in part 3 of the code. The function output and inputs should be changed as follows:[5]

```
function [a0,b0,h] = LLreg(X,y,x,h0,func)
```

The snippet of code below should be substituted for part 3 of Kreg. We first define the results matrices a0 and b0, defined as in Equation 8.16. Then, looping over each point at which we wish to evaluate the function, the design matrix Xtilde and the weight matrix W are created, and manipulated as in Equation 8.17. The only function that you might not have seen before is diag. If you pass diag a vector, as we do below, it will create a square diagonal matrix with the elements of the vector on the main diagonal. If, however, you pass diag a matrix, the function will return a column vector of the main diagonal elements of the matrix.

---

[5] Remember also to update the function help file!

```
1  %----- (3) Perform local linear regression! -----
2  a0 = NaN(n, 1);
3  b0 = NaN(n, 1);
4  for i=1:N
5      Xtilde = [ones(N, 1) (X-x(i))];
6      u = (X - x(i))/h;
7      Ku = krnl(u);
8      W = diag(Ku);
9      bhat = (Xtilde'*W*Xtilde)\(Xtilde'*W*y);
10     a0(i) = bhat(1);
11     b0(i) = bhat(2);
12 end
```

8.5 **Review and Exercises**

Table 8.3 Chapter 8 Commands

Command	Brief Description
class	Returns the name of the class of the object passed to it
date	Returns a string containing the date in dd-mmm-yyyy format
diag	Create diagonal matrix or get diagonal elements of a matrix
fmincon	Routine to minimize an objective function subject to linear or non-linear constraints
fieldnames	Returns structure field names
nargin	Returns the number of input arguments that were used to call the function
strcmpi	Compare strings and cell strings in a case insensitive manner
struct	Creates a structure array with the specified fields and values

This chapter has given you an introduction to kernel based smoothing methods in MATLAB. Good textbook treatment covering the theory and statistical inference of kernel regression estimators include Pagan and Ullah (1999), Yatchew (2003), and Li and Racine (2006). Useful survey articles include Ichimura and Todd (2007), Delgado and Robinson (1992), and Härdle and Linton (1994).

Key early references for kernel density estimation are Rosenblatt (1956) and Parzen (1962), and nonparametric kernel regression was first suggested in Nadaraya (1964) and Watson (1964). In this chapter, we have not discussed statistical inference, although we do ask you to construct pointwise confidence intervals in Exercise (i). In Section 8.4, the local linear estimator was introduced. An alternative standard local regression estimator is the Lowess estimator (locally weighted scatterplot smoothing), which uses a variable bandwidth and downweights outlier observations. The approach is attractive given its better behaviour at the boundary and robustness to outliers. It is covered in Exercise (vii).

For simplicity, we have focused on the use of, and methods to select, a *single* global value for the bandwidth. Some methods allow for variable bandwidth choices: for example, k-nearest neighbour estimation (kNN) and the super-smoother. kNN estimates differ from kernel estimates as they define the size of the neighbourhood according to the observations whose x observations are among the k-nearest neighbours of x_0 in Euclidean distance. The idea was introduced by Loftsgaarden and Queenberry (1965). The question under this approach becomes one of setting the number of neighbours, rather than of the bandwidth. See Exercise (iii) for implementation of the approach in MATLAB. Friedman (1984) supersmoother uses a local cross validation method applied to kNN estimates and is covered in Härdle (1991). Rather than use a fixed k, a variable k is allowed for, determined by local cross validation that entails nine passes over the data.

Series estimators are the other main class of nonparametric estimators not covered in this text. Rather than smoothing over observations, series methods approximate an unknown function with a flexible parametric function, i.e. by a weighted sum of functions. The method thus aims to provide a global approximation to the regression function, as opposed to the local approach of kernel methods. In the series context, the number of parameters to be estimated plays a similar role to the bandwidth in a kernel regression. For real valued x, a common series approximation is the pth order polynomial, i.e. one approximates an unknown function by a sum of polynomials of x, including all powers and cross products where x is vector valued. Another common series approximation is a continuous, piecewise polynomial function known as a spline.

Many estimators have been put forward to estimate the average treatment effect on the treated, including so-called *matching estimators*. Such estimators impute non-treatment outcomes for treated individuals by matching each treated individual to observationally similar untreated individuals. Heckman et al. (1998) develop local polynomial estimators to construct matched outcomes.

EXERCISES

(i) Edit `Kreg` to return also the pointwise 95% confidence interval. *Hint:* The pointwise standard error (ignoring the presence of bias and assuming a homoskedastic error) is constructed as:

$$se(x_0) = \sqrt{\left(\frac{b_K \hat{\sigma}^2}{h \hat{p}(x_0) N} \right)},$$

where $b_K = \int K^2(u)du$ (see Table 8.1) and

$$\hat{\sigma}^2 = \frac{1}{N} \sum_{i=1}^{N} (y_i - \hat{y}_i)^2.$$

(ii) Create a function to return kernel estimates of the *density* of a univariate random variable. *Hint:* the kernel density estimator takes the form:

$$\hat{f}(x_0) = \frac{1}{hN} \sum_{i=1}^{N} K_h(X_i - x_0).$$

(iii) The k-Nearest neighbour estimator is the weighted average of the y values for the k observations closest to x_0. Define $N_k(x_0)$ as the set of k observations closest to x_0. Then,

$$\hat{m}_{kNN}(x_0) = \frac{1}{k} \sum_{i=1}^{N} 1(X_i \in N_k(x_0))y_i.$$

    (a) Create a function kNN to compute a k-Nearest neighbour estimator, computing the distance between observations as the Euclidean distance.

    (b) How does the regression output change as you vary k?

(iv) Check that Equations 8.10 and 8.12 calculate the same value for the sum of squared leave-one-out residuals. Do this by altering parts 3 and 4 of MSE to compute a *literal* leave-one-out estimate of the regression function.

(v) Create a function generalizing the local linear estimator to an n-degree local polynomial estimator. The function should accept inputs and return outputs as:
```
function yhat = LPoly(X, y, x, degree)
```
where degree is an integer in the set $[0, 1, 2, 3]$.

(vi) Repeat the cross validation example of Section 8.3 but with X drawn from a Pareto distribution. Add a trimming function to MSE. How do your estimates change as α (the parameter of the Pareto) increases? How does the cross validation bandwidth change with α and your specification of the trimming function?

(vii) Create a script to implement the LOWESS algorithm, with I iterations.

    (a) Fit a weighted local polynomial regression, using a k-nearest neighbour fitting (see Exercise 8.3).

$$\beta^0 = \arg\min_{\beta} \frac{1}{N} \sum_{i=1}^{N} W_{ki}(x_0) \left(y_i - \sum_{j=0}^{3} \beta_j X_i^j \right)^2$$

    where $W_{ki}(x_0)$ is the k-NN weight.

    (b) Compute the residuals, $\hat{\varepsilon}_i$ and $\hat{\sigma} = \text{median}(\hat{\varepsilon}_i)$.

    (c) Calculate robustness weights $\delta_i = K\left(\hat{\varepsilon}_i/6hat\sigma\right)$, where $K(u)$ is the Quartic kernel.

    (d) Re-run regression with weights $\delta_i W_{ki}(x_0)$.

    (e) Repeat steps 2 and 3 a further $I - 1$ times.

9 Semiparametric Methods

Variatio semper delectat.

Phaedrus Augusti Libertus[*]

Variation always delights—well, at least in the arena of semiparametric regression! In practice, nonparametric regression is rarely used in economics for anything more than data description and univariate estimation. This stems from the economist's desire to model multivariate relationships. The statistical precision of nonparametric estimators decreases a lot when modelling multiple explanatory variables. Intuitively, for high-dimensional settings, there are often too few observations in the local neighbourhood of a particular point. This causes nonparametric methods to break down and is often referred to as the 'Curse of Dimensionality'.

As a further limitation, the output of high-dimensional nonparametric regression is too much for most people to get their heads around—output visualization gets difficult as soon as the dimensionality of X exceeds two. An infinite-dimensional object that you cannot visualize is typically a rather useless beast.

Given these problems, models have been developed that reduce the complexity of high dimensional regression problems, making them better suited to nonparametric estimation. This often involves an allowance for partly parametric modelling. These so-called 'semiparametric' models are useful as they can provide a better balance of efficiency and flexibility, and yield output that is easier to interpret. The estimation of semiparametric models often requires the use of nonparametric techniques—so, your efforts in Chapter 8 were not in vain!

Given the myriad ways in which you might combine parametric and nonparametric components, it should not be surprising that there are a large number of semiparametric estimators available. In this chapter, we will guide you through the estimation of two popular models: the Partially Linear Model and the Single Index Model. Before doing so, we will return briefly to the case of pure nonparametric regression to introduce multivariate kernel regression and to highlight its difficulties.

---

[*] Phaedrus. (1745). *The fables of Phaedrus: translated into English prose, as near the original as the different idioms of the Latin and English languages will allow*. London: Joseph Davidson.

9.1 **Multivariate Kernel Regression**

There are no fundamental methodological differences between kernel regression in the univariate and multivariate settings. In both, we attempt to approximate the value of the response curve at **x** by the local average of response variables in the close neighbourhood of **x**. In the multidimensional setting, the local neighbourhood of **x** is a multidimensional ball. Rather than a single dimension and a single bandwidth, there are now multiple dimensions and, potentially, multiple bandwidths.

The multivariate generalization of the NW estimator takes the following form, where $K_H(\mathbf{X}_i - \mathbf{x})$ is a multivariate kernel function:

$$\hat{m}_H(\mathbf{x}) = \frac{\sum_{i=1}^{N} K_H(\mathbf{X}_i - \mathbf{x}_0)y_i}{\sum_{i=1}^{N} K_H(\mathbf{X}_i - \mathbf{x}_0)} \tag{9.1}$$

$$= \frac{\sum_{i=1}^{N} K\left(\frac{X_i^1 - x_0^1}{h^1}, \ldots, \frac{X_i^K - x_0^K}{h^K}\right)y_i}{\sum_{i=1}^{N} K\left(\frac{X_i^1 - x_0^1}{h^1}, \ldots, \frac{X_i^K - x_0^K}{h^K}\right)}. \tag{9.2}$$

The multivariate kernel function K_H is typically chosen as the product of univariate kernels:

$$K_H(\mathbf{u}) = \prod_{j=1}^{K} K_{h_{-j}}(u_j) \tag{9.3}$$

If a product kernel is used, it is thought to be good practice to transform the regressors to a common scale by dividing by the standard deviation, or to use multiple bandwidths.

The function `BiKreg` performs the regression of y on a bivariate **X**. While similar to the function `Kreg` (which was introduced in Chapter 8), there are a few differences that make it suitable for the bivariate case. The function is designed to accept the data and an $n \times 2$ matrix, x, that is used to create the grid on which the function is evaluated. On line 20, we unpack the two columns of x and create an $n \times n$ grid of evaluation points using `meshgrid`. Parts 2 and 3 of the function follow similarly to the univariate case, except for some minor changes when creating the bandwidth (to ensure conformability of the matrices). In the interests of space, only the Gaussian kernel is used in `BiKreg`; if you wish, you can easily edit `BiKreg` to allow for different kernels as we did in `Kreg`.

For ease of understanding, in the final part of the function, the product kernels associated with each variable are generated separately, before being bought together in the final loop of the code. This is not the neatest way of coding the estimator because of all the loops (see Chapter 10), but we present it in this way to provide an intuitive introduction to the concept. How could you refine the code?

BiKreg.m

```matlab
function [yhat,h,w,z] = BiKreg(X,y,x,h0)
%-------------------------------------------------
% PURPOSE: performs the kernel regression of y
%          on (bivar) X
%-------------------------------------------------
% USAGE:  [yhat,h,w,z] = BiKreg(X,y,x,h0)
% where:  y  : N-by-1 dependent variable
%         X  : N-by-2 independent variable
%         x  : n-by-2 points of evaluation
%         h0 : 2-by-1 bandwidth
%-------------------------------------------------
% OUTPUT: h : bandwidths used
%         yhat : regression evaluated at the
%                n-by-n grid from columns of x
%-------------------------------------------------

%--- (1) Create grid of points ----------------
N = length(y);
n = size(x, 1);
[w,z] = meshgrid(x(:,1), x(:,2));


%--- (2) Set bandwidth if not supplied & kernel--
if nargin < 4
    hx = median(abs(X-repmat(median(X), ...
        N,1)))/0.6745*(4/3/N)^0.2;
    hy = median(abs(y-repmat(median(y), ...
        N,1)))/0.6745*(4/3/N)^0.2;
    h = sqrt(hy.*hx);
else
    h = h0;
end
hw = h(1);
hz = h(2);
krnl = @(u) exp(-(u.*u)/2)/sqrt(2*pi);

```

```
38   %--- (3) Perform bivariate kernel regression!---
39   % Product kernel X(:,1)
40   KW = NaN(length(y),size(w,1));
41   W = X(:,1);
42   for w_  = 1 : size(w,1)
43       wi = w(1,w_);
44       uw = (W - wi)/hw;
45       Kuw = krnl(uw);
46       KW(:,w_)  = Kuw;
47   end
48
49   % Product kernel X(:,2)
50   KZ = NaN(length(y),size(z,1));
51   Z = X(:,2);
52   for z_  = 1 : size(z,1)
53       zi = z(z_, 1);
54       uz = (Z - zi)/hz;
55       Kuz = krnl(uz);
56       KZ(:,z_)  = Kuz;
57   end
58
59   % NW estimator
60   yhat = NaN(size(w,1), size(w,2));
61   for i = 1 : n
62       for j = 1 : n
63           Ku = KW(:,i).*KZ(:,j);
64           yhat(j,i)  = sum(Ku.*y)/sum(Ku);
65       end
66   end
67
68   return
```

To see how this works in practice, let's simulate the following data generating process and plot the output.

$$W_i \sim U(0,2) \tag{9.4}$$

$$Z_i \sim U(0,5) \tag{9.5}$$

$$\varepsilon_i \sim \mathcal{N}(0,1) \tag{9.6}$$

$$y_i = \sin(\pi W_i) \cdot \cos(Z_i) + \Phi(X_i Z_i) + \varepsilon_i \tag{9.7}$$

This is done in Matlab as:

BiKregScript.m

```matlab
% create data
W = 2*rand(1000,1);
Z = 5*rand(1000,1);
X = [W Z];
y = sin(pi*W).*cos(Z) + normcdf(W.*Z) +...
    rand(length(W),1);
x = [linspace(0, 2, 100)' linspace(0, 5, 100)'];

% bivariate kreg
[yhat,h,w,z] = BiKreg(X,y,x);

surf(w,z,yhat)
xlabel('W', 'FontSize', 14);
ylabel('Z', 'FontSize', 14);
zlabel('y', 'FontSize', 14);
hold on
scatter3(W,Z,y)
```

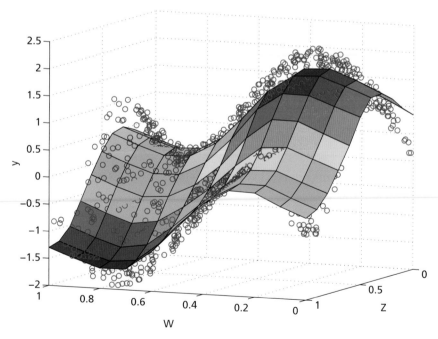

Figure 9.1 Multivariate Kernel Regression

Nonparametric regression techniques can just about manage in the bivariate case (given sufficient data). However, in higher dimensions, nonparametric regression faces some rather serious practical problems. Most importantly, it is afflicted by the 'Curse of Dimensionality'; the rate of convergence of nonparametric estimates to the truth falls in the number of dimensions of the problem. Further, for the applied researcher, plotting high dimensional outputs is rarely possible, making the interpretation of results very difficult. (It is not clear how we could easily add another dimension to Figure 9.1 . . .) For this reason, we now move on to explore some key semiparametric methods.

9.2 **Dimension Reduction**

Models have been developed that reduce the complexity of high dimensional regression problems. This often involves an allowance for partly parametric modelling, enabling us to reduce the dimension of the regression problem, while still allowing for some nonparametric relationships. There are three main approaches that you might choose to follow:

(i) Variable selection;
(ii) Use of semiparametric or nonparametric indices;
(iii) Use of nonparametric link functions.

VARIABLE SELECTION

To reduce the dimension of the regression problem, you can always just include fewer variables! There is no rule of thumb for selecting a subset of variables, although looking to economic theory often provides useful modelling insights. However, when selecting variables, you must keep in mind the impact that this has on the explanatory power of the model—although fewer variables might help to alleviate the Curse of Dimensionality, you might end up omitting key data.

NONPARAMETRIC INDICES

Rather than drop a subset of variables, it is sometimes possible to partition the explanatory variables into one set to be modelled nonparametrically and another set to be modelled parametrically. It is common, for instance, to model certain relationships linearly, while allowing other relationships to be determined by the data. For example, when modelling the relationship between income and the budget share devoted to alcohol, you might want to control for household composition, education, and other demographic variables, while allowing for a flexible relationship between income and alcohol expenditure. In this context, you might want to estimate the model:

$$w_i = g(m_i, X_i) + \varepsilon_i \tag{9.8}$$

$$= h(m_i) + \beta X_i + \varepsilon_i, \tag{9.9}$$

where m gives income, and X is a matrix of household composition, education, and other demographic variables. This is called a Partially Linear Model. In Section 9.3, we will code up the estimator proposed by Robinson (1988) for this model.

NONPARAMETRIC LINK FUNCTIONS

An index summarizes the values of many different variables into a single number. For example, the inflation rate is a price index; it combines information on the price changes occurring for different goods in different places into a single number. The dimensionality of a problem is greatly reduced (to one!) if the information contained by multiple variables is reduced to a single index. In Section 9.4, we will code the Semiparametric Least Squares estimator, proposed by Ichimura (1993), for this model.

9.3 Partially Linear Models

Partially Linear Models are popular for applied work. They provide a simple way of incorporating a nonparametric component into an otherwise linear model. The model has two parts: a parametric component $(X\beta)$ and a nonparametric component $(m(Z))$:

$$y_i = X_i\beta + m(Z_i) + \varepsilon_i \tag{9.10}$$

The goal is to estimate β and $m(\cdot)$. We will restrict attention to the case where Z_i is unidimensional. The extension to a multivariate nonparametric component is straightforward now that you are familiar with multivariate kernels.

It is worth noting that simply regressing y on X will return inconsistent estimates for β, except in the unlikely case where $\text{Cov}(X, m(Z)) = 0$—so do not do this! Further, since $m(\cdot)$ is unconstrained, X_i cannot be perfectly collinear with any smooth function of Z_i. This means that an intercept and any deterministic function of Z_i must be excluded from X_i, as the function $m(\cdot)$ will embody these components.

9.3.1 ROBINSON'S APPROACH

We will estimate the model following the seminal approach of Robinson (1988). Robinson's estimator involves 'concentrating out' the unknown function using a double residual regression to first estimate β. The resulting estimator for β is consistent, asympotically normal, and converges at the parametric rate. We will then use the estimated $\hat{\beta}$ to recover $\hat{m}(\cdot)$.

To see the essential intuition behind the Robinson estimator, take expectations of the Partially Linear Model with respect to Z_i:

$$\mathbb{E}(y_i|Z_i) = \mathbb{E}(X_i\beta|Z_i) + \mathbb{E}(m(Z_i)|Z_i) + \mathbb{E}(\varepsilon_i|Z_i) \qquad (9.11)$$

$$= \mathbb{E}(X_i|Z_i)\beta + m(Z_i). \qquad (9.12)$$

Subtracting this from the full model removes the unknown function $m(\cdot)$:

$$y_i - \mathbb{E}(y_i|Z_i) = [X_i - \mathbb{E}(X_i|Z_i)] \cdot \beta + \varepsilon_i. \qquad (9.13)$$

This expression can be given as a relationship between residuals as:

$$\varepsilon_{yi} = \varepsilon_{xi}\beta + \varepsilon_i. \qquad (9.14)$$

β can therefore be recovered using a regression of conditional errors.

Robinson (1988) suggested estimating $\mathbb{E}(y_i|Z_i)$ and $\mathbb{E}(X_i|Z_i)$ by standard kernel regressions and then using the residuals $\hat{\varepsilon}_{yi}$ and $\hat{\varepsilon}_{xi}$ to estimate $\hat{\beta}$. Formally, let

$$y_i = m_y(Z_i) + \varepsilon_{yi} \qquad (9.15)$$

$$X_i = m_x(Z_i) + \varepsilon_{xi}, \qquad (9.16)$$

with the estimators

$$\hat{m}_y(z) = \frac{\sum_{i=1}^{N} K_h(z - Z_i)y_i}{\sum_{i=1}^{N} K_h(z - Z_i)} \qquad (9.17)$$

$$\hat{m}_x^k(z) = \frac{\sum_{i=1}^{N} K_h(z - Z_i)X_i^k}{\sum_{i=1}^{N} K_h(z - Z_i)}, \qquad (9.18)$$

where $\hat{m}_x^k(z)$ gives the estimator of the conditional mean of the kth dimension of X given Z. (It might be appropriate to also trim the data, such that certain observations are downweighted, in the standard way discussed in Section 8.3.1.)

The estimator of β is then obtained as:

$$\hat{\beta} = \left[\sum_{i=1}^{N}(X_i - \hat{X}_i)'(X_i - \hat{X}_i)\right]^{-1} \sum_{i=1}^{N}(X_i - \hat{X}_i)'(y_i - \hat{y}_i) \qquad (9.19)$$

where $\hat{y}_i = \hat{m}_y(z)$ and $\hat{X}_i = \hat{m}_x(Z_i)$.

Implementation of the Robinson estimator requires the regression of y_i and each of the separate dimensions of X_i on Z_i. These should be viewed as separate kernel regressions, with different bandwidths if required. This is easily done using the techniques introduced in Chapter 8. As an example, consider the data generating process:

$$y_i = X_{i,1}\beta_1 + X_{i,2}\beta_2 + \sin(8Z_i) + \varepsilon_i \qquad (9.20)$$

where $X_1 \sim \mathcal{N}(1, 4)$, $X_2 \sim \mathcal{N}(1, 0.25)$, $Z \sim U(0, 1)$, and $\varepsilon \sim \mathcal{N}(0, 1)$.

The code below generates 1,000 observations from the data generating process with a coefficient vector $\beta = [2, -3]$. We then proceed, in parts 2 and 3 of the code, to concentrate out the nonparametric component of the model. First using a nonparametric regression of y on Z (saved as ehatY) and then regressing each separate dimension of X on Z (saved as ehatX). We here make use of the function Kreg that was introduced in Chapter 8.

PLMScript.m

```
1  %----- (1) Simulate data  ----------------------
2  X1 = 1 + 2*randn(1000,1);
3  X2 = 1 + 0.5*randn(1000,1);
4  X = [X1 X2];
5  Z = rand(1000, 1);
6
7  beta = [2;-3]; % make up a beta
8  y = sin(8*Z) + X*beta + randn(length(X),1);
9
10 %----- (2) NW regression of y on Z -----------
11 ghat_y = Kreg(Z,y,Z);
12 ehatY = y - ghat_y;
13
14 %----- (3) NW regression of X on Z -----------
15 ehatX = NaN(size(X,1),size(X,2));
16 for i = 1:size(X,2)
17     ghat_X = Kreg(Z,X(:,i),Z);
18     ehatX(:,i) = X(:,i) - ghat_X;
19 end
```

Finally, the estimated coefficient vector bhat is returned by regressing the residuals on each other.

```
>> bhat = (ehatX'*ehatX)\(ehatX'*ehatY)

bhat =
```

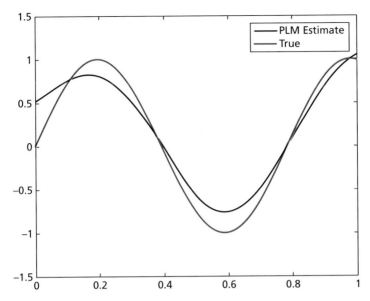

Figure 9.2 Nonparametric Component of PLM

```
     1.9982
    -2.9217
```

The estimator has performed well. The nonparametric component of the model, $m(Z)$, is recovered easily using the estimated coefficient vector. Since $m_z(Z_i) = \mathbb{E}(y_i - X_i\beta | Z_i)$, a consistent estimator of $m_z(Z_i)$ is given by:

$$\hat{m}_y(z) = \frac{\sum_{i=1}^{N} K_h(z - Z_i)(y_i - X_i\hat{\beta})}{\sum_{i=1}^{N} K_h(z - Z_i)}. \tag{9.21}$$

The following code uses the estimated coefficient vector to recover the nonparametric component of the model and plots the output. We can see that the shape of the nonparametric function is well approximated by the nonparametric estimate in Figure 9.2.

```
1  %----- (4) NW regression of (y - X*beta) on Z ---
2  y_bhat = y - X*bhat;
3  [y_bhat, order] = sort(y_bhat, 'ascend');
4  Z_ = Z(order);
5  z_ = min(Z_):(max(Z_) - min(Z_))/100:max(Z_);
6  [mhat,h] = Kreg(Z_,y_bhat,z_, 0.08);
7
8  plot(z_,mhat,'k',z_,sin(8*z_),'b','LineWidth',2)
```

9.4 **Single Index Models**

Single index models (SIMs) are another important and popular class of semi-parametric methods. In these models, $\mathbb{E}(y|X) = g(X\boldsymbol{\beta})$, where $g(\cdot)$ is often referred to as the link function. These models are only nonparametric in one dimension—which is very useful for alleviating the Curse!

You will have come across SIMs many times before—even in the course of this book! Most parametric models are single index—Normal regression, Logit, Probit, Tobit.... These models are based on strong functional form and distributional assumptions, which you might want to relax—or at least examine the robustness of your results to relaxing these restrictions.

Semiparametric SIMs keep a linear form for the index $(X\boldsymbol{\beta})$ but allow the link function $(g(\cdot))$ to be any smooth function. The challenge is to find an estimator for $\boldsymbol{\beta}$ that attains the $\sqrt{N}$-rate of convergence. Two different approaches are typically taken to estimate $\boldsymbol{\beta}$ and $g(\cdot)$:

(i) Iterative approximation of $\boldsymbol{\beta}$ using Semiparametric Least Squares (SLS) or pseudo Maximum Likelihood estimation (PMLE).
(ii) Direct estimation of $\boldsymbol{\beta}$ using the average derivative of the regression function.

Although the nitty-gritty varies across the different methods, the basic approach to estimation can be summarized as follows:

(i) Estimate $\boldsymbol{\beta}$ by $\hat{\boldsymbol{\beta}}$.
(ii) Compute index values $\hat{v} = X\hat{\boldsymbol{\beta}}$.
(iii) Estimate the link function $g(\cdot)$ using nonparametric regression of y on $\hat{v}$.

In the rest of this chapter, we will guide you through the implementation of Ichimura's (1993) Semiparametric Least Squares estimator, which falls under the first approach. In Exercises (iii) and (iv), you are walked through a Pseudo Maximum Likelihood Estimator and an Average Derivative Estimator.

Before getting started, note that X cannot include an intercept—the function $g(\cdot)$ will include any location and level shift. Further, the level of $\boldsymbol{\beta}$ is not identified and thus a normalization criterion is required. This is typically achieved by setting one element of $\boldsymbol{\beta}$ equal to 1. In this case, you must be careful that this variable correctly has a non-zero coefficient. Identification also requires X to include a continuously distributed variable that is informative for y—you cannot identify the continuous function m on a discrete support.

9.4.1 SEMIPARAMETRIC LEAST SQUARES

Semiparametric Least Squares was introduced by Ichimura (1993). The approach is motivated by the insight that, if $g(\cdot)$ were known, we could estimate

β using non-linear least squares. This would involve picking β to minimize the criterion function:

$$S(\beta, g) = \sum_{i=1}^{N} (y_i - g(X_i\beta))^2 \qquad (9.22)$$

Sadly for us, this estimator is infeasible—we do not know the structure of the nonparametric link function (this is what we are trying to find out!).

Ichimura proposed replacing the unknown function $g(\cdot)$ with the 'leave-one-out estimator', resulting in a technique with a similar flavour to the cross validation method (which was covered in Chapter 8). In fact, Härdle, Hall, and Ichimura (1993) suggest picking β and h to jointly minimize $S(\beta)$. Under the approach, the parameter vector β is chosen to minimize:

$$S(\beta) = \sum_{i=1}^{N} \pi_i(y_i - \hat{g}_{\neg i}(X_i\beta))^2 \qquad (9.23)$$

where

$$\hat{g}_{\neg i}(X_i\beta) = \frac{\sum_{i \neq j} K\left(\frac{(X_i - X_j)\beta}{h}\right) y_i}{\sum_{i \neq j} K\left(\frac{(X_i - X_j)\beta}{h}\right) y_i} \qquad (9.24)$$

and π_i is a trimming function, which downweights or drops observations if $X_i\beta$ is too small (see Section 8.3.1). For heteroskedastic data, one can also easily incorporate a weight function.

Casting your mind back to Chapter 8, you will remember that, luckily, the leave-one-out method is not as computationally intensive as it might first appear. It can be shown that:

$$y_i - \hat{g}_{\neg i}(X_i\beta) = \frac{y_i - \hat{g}(X_i\beta)}{1 - K_h((X_i - X_i)\beta) / \sum_i K_h((X_i - X_i)\beta)} \qquad (9.25)$$

Therefore, the value of the leave-one-out objective function $S(\beta)$, for given choices of h and g, requires only a single computation of the regression function.

Implementation of the Ichimura estimator brings us back (yet again!) to our old friend `fmincon`. Inspecting Equations 9.23 and 9.24 highlights that there are two elements to the estimation process for β:

(i) For a given β and h, evaluate $S(\beta)$;
(ii) Jointly select the β and h that minimize $S(\beta)$.

To conduct the first part of the procedure, we code the function `MSEg` that returns the mean squared error associated with any supplied bandwidth and parameter vector. To optimize our choice over these parameters, this function is then passed to `fmincon`.

To begin, simulate a data generating process, which we will then try to recover using the Ichimura estimator. As the example here, we'll simulate the process below. However, feel free to experiment!

$$y_i = (\mathbf{X}\boldsymbol{\beta})^2(1 + \sin(\mathbf{X}\boldsymbol{\beta}))\varepsilon_i \qquad (9.26)$$

where $X_1, X_2 \sim U(-1, 1)$ and $\varepsilon \sim \mathcal{N}(0, 1)$,

Specifying $\boldsymbol{\beta} = [1, 3]'$, generate 1,000 observations drawn from the data generating process:

```
1  %----- Set Parameters & DGP ----------------
2  N = 1000;
3  X = -1 + 2*rand(N,2);
4  beta = [1;3];
5  Xbeta = X*beta;
6  y = Xbeta.^2.*(1 + sin(Xbeta)) + randn(N,1);
```

Our task is to recover $\boldsymbol{\beta}$ and the link function from the simulated data. Below, we define the function MSEg that calculates the value of the objective function for any choice of the bandwidth and $\boldsymbol{\beta}$. The first two inputs are simply the data vectors, the final input is a candidate parameter vector. The first element of param is the bandwidth for the nonparametric regression, the remaining elements give the coefficient vector $\boldsymbol{\beta}$. The function starts by defining new variables h and beta as the relevant elements of param. An anonymous Gaussian kernel function is then defined on line 20 for use in the estimation of the nonparametric link function.

The action happens in part 3 of the code. Using the supplied values for h and $\boldsymbol{\beta}$, the function evaluates the kernel weights and regression function for the values of the independent variable at each observation, storing the important bits of the output in the summary variables ghat, weighti and weightall. Lines 37 and 38 calculate the value of Equation 9.25 to evaluate $S(\boldsymbol{\beta})$, given the supplied parameter vector.

MSEg.m

```
1  function Sbeta = MSEg(X, y, param)
2  %--------------------------------------------
3  % PURPOSE: implements Ichimura's SIM estimator
4  % imposing beta(1) = 1 as the normalization
5  % condition
6  %--------------------------------------------
7  % INPUTS: y : n-by-1 dependent variable
8  %         X : n-by-K independent variable
9  %         param : (1+K)-by-1 vector of
```

```
10  %                    parameters (bandwidth and beta)
11  %-------------------------------------------------
12  % OUTPUT: Sbeta : MSE of leave one out estimator
13  %-------------------------------------------------
14  h=param(1) ;
15  beta=param(2:end) ;
16  N = length(y);
17  Xbeta = X*beta;
18
19  %----- (1) Declare kernel ---------------------
20  krnl = @(u) exp(-(u.*u)/2)/sqrt(2*pi);
21
22  %----- (2) Declare summary vectors ------------
23  yhat = NaN(N,1);
24  weight_i = NaN(N,1);
25  weight_sum = NaN(N,1);
26
27  %----- (3) NW regression & save weights --------
28  for i = 1:N
29      u = (Xbeta - Xbeta(i))/h;
30      Ku = krnl(u);
31      weight_i(i) = Ku(i);
32      weight_sum(i) = sum(Ku);
33      yhat(i) = sum(Ku.*y)/sum(Ku);
34  end
35
36  %----- (4) Calculate leave-1-out MSE -----------
37  residual=(y - yhat)./(1-(weight_i./weight_sum));
38  Sbeta = sum(residual.^2);
39
40  return
```

With MSEg defined, all we need now are the starting values and constraints for the optimization procedure. The following section of code begins by estimating starting values for β and h to pass to fmincon using OLS and the plug-in bandwidth. We then build the constraint equality matrices, Aeq and beq, to enforce the identification condition on β—namely, $\beta_1 = 1$. The elements of the constrained optimization problem are then passed to fmincon where the parameter vector is selected to minimize the leave-one-out mean squared error using the function MSEg.

The final step of the estimation procedure is the recovery of the link function. Using the estimated parameter vector, the explanatory variables are transformed to Xbeta, which, alongside the optimal bandwidth opth, is passed to the kernel regression function Kreg that was constructed in Chapter 8.

SIMScript.m

```matlab
%----- (1) Starting values --------------------
y_ = y - X(:,1);
X_ = X(:,2:end);
beta0 = (X_'*X_)\X_'*y_;
beta0 = [1; beta0];

% starting values for the h's (Bowman and
% Azzalini (1997))
x_ = X*beta0;
hx=median(abs(x_-median(x_)))/0.6745*(4/3/N)^0.2;
hy=median(abs(y-median(y)))/0.6745*(4/3/N)^0.2;
h=sqrt(hy*hx);

% param starting value
param0 = [h;beta0];

%----- (2) Optimal constraints -----------------
Aeq = zeros(1, length(param0));
Aeq(2) = 1;
beq = 1;

% lower boound on h
lb = [0;-Inf(length(beta0),1)];

%----- (3) Optimize beta and h -----------------
options = optimset('Display', 'off',...
    'Algorithm', 'sqp', 'MaxFunEvals', 1e5,...
    'MaxIter', 1e5,'TolX',1e-10, 'TolFun',1e-10);
Param = fmincon(@(p) MSEg(X, y, p), param0,...
    [], [], Aeq,beq,lb, [], [] ,options);

opth = Param(1);
optbeta = Param(2:end);

%----- (4) Recover link function ----------------
Xbeta = X*optbeta;
xbeta = [-4:0.1:4]';
ghat = Kreg(Xbeta, y, xbeta, opth);
```

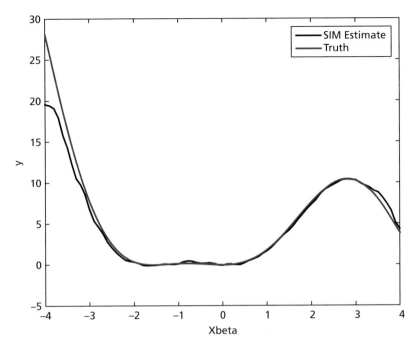

Figure 9.3 Recovered Link Function

Eye-balling the output, we can see that the procedure does a good job! Figure 9.3 shows the nonparametric link function recovered from the data— the wiggles in the index function are picked up by the procedure. The returned coefficient vector is accurately recovered.

```
>> optbeta = Param(2:end)

optbeta =

    1.0000
    3.0874
```

How would you assess the accuracy of the output if $\beta_1 \neq 1$?

When implementing this estimator in your own work, we recommend that you check the sensitivity of your estimates to the choice of starting value and the tolerance values passed to fmincon. $S(\beta)$ may be non-convex and multimodal, making the results potentially sensitive to the settings used by the MATLAB optimization routines.

9.5 **Review and Exercises**

Table 9.1 Chapter 9 Commands

Command	Brief description
scatter3	3D scatter plot
setdiff	Set difference between first and second inputs
surf	Plot 3D coloured surface

9.5.1 FURTHER READING

For a comprehensive introduction to semiparametric regression, see the textbooks of Härdle et al. (2004), Pagan and Ullah (1999), and Yatchew (2003). Useful surveys for semiparametric methods are given by Delgado and Robinson (1992), Ichimura and Todd (2007), and Powell (1994).

Single Index Models (SIMs) were partially covered in this chapter. There are two main approaches taken to estimate SIMs: iterative approximation and average derivative estimation. We introduced you to Semiparametric Least Squares estimator as an example of the iterative approximation approach. When the dependent variable is discrete, *Pseudo Maximum Likelihood* estimation is an alternative to this approach. In Exercise (iii), we ask you to code the Klein and Spady (1993) PML estimator. Manski (1975) and Horowitz (1992) provide Maximum Score and Smoothed Maximum Score estimator for semiparametric binary choice models.

An alternative, direct approach to SIM estimation is the *Average Derivative Estimator* that was first proposed by Stoker (1986) and extended to the weighted case by Powell et al. (1989). The basic intuition comes from the fact that, if the conditional mean is a single index, then the vector of average derivatives of the conditional mean determines β up to a scale. Exercise 9 (iv) walks you through the implementation of the Stoker (1986) estimator.

Additive Models are another important class of semiparametric estimators. They assume separability of the regressors but are otherwise nonparametric: $y = g_1(x_1) + g_2(x_2) + \ldots + g_K(x_K)$. These models can be solved by *backfitting* algorithms (see Hastie and Tibshirani (1990)). Plots of the estimated subfunctions give the marginal effect of the variable of interest. Exercise (v) walks you through backfitting for additive models. Additive models have been generalized to allow for a nonparametric link function. Hastie and Tibshirani (1990) describe how to modify backfitting procedures to allow for binary response and survival data. For unknown link functions, see the Alternating Conditional Expectation procedure of Linton et al. (1997).

The label 'semiparametric' is also applied to models that are, for example, linear in parameters but have a flexible model of the conditional variance— that is, $\text{Var}(y|X) = \sigma^2(X)$. Indeed, it is in context that the development of semiparametric methods can be traced back to, and also for the identification and estimation of censored regression and sample selection models. See Powell (1994) for an overview.

EXERCISES

(i) Simulate a Gaussian mixture model, with four Gaussian distributions. Estimate the density and use the MATLAB function `contour` to plot the contours of the *pdf*. *Hint:* You have not yet got a bivariate kernel density estimator coded up—see Exercise 8.1 for insights for the univariate case.

(ii) Using a Quartic kernel, adapt the code for the Ichimura (1993) SLS estimator to also return the marginal effects.

(iii) Construct a function for the Klein and Spady estimator (1993)—a PML method for binary response data. Note that, if the distribution of the error term was known, the classical, MLE would maximize:

$$L_N(\boldsymbol{\beta}) = \sum_{i=1}^{N} y_i \log\left(F(\mathbf{X}_i\boldsymbol{\beta})\right) + (1 - y_i) \log\left(F(\mathbf{X}_i\boldsymbol{\beta})\right).$$

    (a) Obtain starting values for $\boldsymbol{\beta}$ using a logit regression (see Chapter 4).

    (b) Construct the constraint matrices A and b to impose the condition $\hat{\beta}^1 = 1$.

    (c) Create a function that calculates $\hat{F}(\mathbf{X}_i\boldsymbol{\beta})$ for any given $\boldsymbol{\beta}$, and then constructs the negative of $\hat{L}_N(\boldsymbol{\beta})$. This is your loss function.

    (d) Minimize the loss function subject to the constraint function using `fmincon`.

    (e) Amend your function to also calculate marginal effects.

(iv) For a SIM, it can be shown that the vector of average derivatives of the conditional mean determines $\boldsymbol{\beta}$ up to a scale:

$$\delta = \mathbb{E}\left[g'(\mathbf{X}\boldsymbol{\beta})\right]\boldsymbol{\beta}$$

where $g'(\mathbf{X}\boldsymbol{\beta})$ is a scalar. This motivates the Weighted Average Derivative (WAD) Estimator. Choosing to weight observations by their density, the density weighted average derivative estimator is given as:

$$\hat{\delta} = -\frac{1}{N} \sum_{i=1}^{N} y_i f'(\mathbf{X}) \tag{9.27}$$

Create a function that returns the WAD estimate, using a Quartic kernel.

(v) Build a function to estimate

$$y_i = \beta_0 + \cos(X_{1i}) + X_{2i}^3,$$

where $X_1, X_2 \sim U(0, 20)$, by backfitting. Note that the general model is given as: $y_i = \beta_0 + \sum_{k=1}^{K} f_k(X_{ki}) + \varepsilon_i$, and so:

$$\mathbb{E}\left(y_i = \beta_0 + \sum_{k \neq j} f_k(X_{ki})|X_{ji}\right) = f_j(X_{ji}).$$

Using backfitting, we can estimate the additive functions iteratively.

(a) Run an OLS regression of y on **X** and a constant to generate starting values.
(b) For each $j = 1, \ldots, K$, set:

$$f_k = \hat{\beta}_0 + \sum_{k \neq j} f_k.$$

Estimate f_k by kernel regression.
(c) Iterate. Repeat until the differences in f_k between iterations are sufficiently small (using the `while` command).

Part V
Speed

10 Speeding Things Up . . .

> I have made this longer than usual because I have not had time to make it shorter.
>
> Blaise Pascal*

10.1 Introduction

So far, we have concentrated on writing intuitive code, rather than fast code. One of the beauties of the MATLAB language is that there are many 'low hanging fruits' waiting to be picked in terms of writing code which runs quickly on our machines. In this final chapter, we discuss a number of tricks to speed things up.

Of course, there is a trade-off involved in these sorts of coding decisions. We could spend a lot of time trying to save time, or we may be happy to write somewhat less optimal code, save our own time, and accept that our computational resources will be tied up for longer.[1] For this reason, this chapter will focus on the quick and easy ways to speed up code in MATLAB. We will also point you in the direction of further resources in case you are interested in spending more time writing *very* quick code . . . .

10.2 Clever Coding

The simplest way to speed up code is to follow a number of basic rules. These rules are almost costless to implement; you should follow them wherever possible.

---

* Pascal, B., *Provincial Letters: Letter XVI* (1656).

[1] Of course, the amount we decide to optimize our code should depend on how frequently we will run a particular program. If we are writing a program to run once or a small number of times it seems unlikely that we would want to spend a long time thinking about speeding up code, whereas if it is a program we will use many times, or repeat many times in a given project, we may be interested in thinking about speeding things up!

10.2.1 VECTORIZING AND PREALLOCATING

Rule 1: Thou shalt vectorize. MATLAB has been written to operate very efficiently on matrices, with the incorporation of a number of clever linear algebra routines to speed up computation. For this reason, you should avoid loops in MATLAB wherever possible. Never operate on a matrix element-by-element if you can operate on the entire matrix at once.

Rule 2: If thou must loop, thou shalt preallocate. You will have noticed throughout this book that we often initialize our matrices before we populate them in a loop. Typically, we have done this by creating matrices of the required dimension with each element being NaN. This may seem very strange; after all, why not just have MATLAB add a new row to a matrix for each iteration of a loop? The answer is simple: it wastes time. First, it wastes time to change the dimensions of a matrix—if possible, we should choose the dimensions of a matrix and then leave them fixed. Second, a matrix created in a piecemeal fashion is likely to be stored in non-contiguous memory. That is, it is likely to be stored in many different places in memory. This will slow subsequent operations on the matrix.

To illustrate the benefits of these two rules, the function `TimeTests` creates a matrix in four different ways. The function makes use of MATLAB's `tic` and `toc` commands to time the various operations—try them out at the command line if you are interested in seeing how they behave.

TimeTests.m

```
 1  function[ratio1, ratio2, ratio3] = TimeTests(n)
 2  %-------------------------------------------------
 3  % PURPOSE: Test performance of loops vs
 4  %              vectorized code
 5  %-------------------------------------------------
 6  % INPUTS: n : The size of the vector that should
 7  %              be filled
 8  %-------------------------------------------------
 9  % OUTPUT: ratio1 : ratio of using a row loop vs
10  %                    vectorizing
11  %          ratio2 : ratio of using a column loop
12  %                    vs vectorizing
13  %          ratio3 : ratio of a preallocated loop
14  %                    vs vectorizing
15  %-------------------------------------------------
16
17  %----- (1) Generate n x 1 vector of ones in 4
18  %              different ways --
```

```
19  tic; vectones = ones(n,1); vect=toc;

20

21  tic; for i=1:n
22        rowloopones(1,i) = 1;
23  end; rowloop=toc;

24

25  tic; for i=1:n
26        loopones(i) = 1;
27  end; loop=toc;

28

29  preloop = NaN(n,1);
30  tic; for i=1:n
31        preloop(i) = 1;
32  end; prealloc = toc;

33

34  %----- (2) Calculate ratio of slow methods to
35  %              fast method -----
36  ratio1 = rowloop/vect;
37  ratio2 = loop/vect;
38  ratio3 = prealloc/vect;
39  return
```

TimeTests builds an N-dimensional vector of ones in four different ways: by directly creating an $N \times 1$ vector (the vectorized way), by looping through a row vector, by looping through a column vector, and by looping once we have first preallocated in two different ways, and finally by looping through a vector where we have.

Let's run this function for a large value of N, say $N = 1,000,000$:

```
>> [r1,r2,r3]=TimeTests(1000000);
>> [r1,r2,r3]

ans =

    90.9239    80.0123    22.9465
```

Of course, this is a relatively contrived example. However, TimeTests still demonstrates an important point. We can speed up our code by a factor of 80–90 if we use vectorization rather than naive loops or by a factor of ~20 times when compared to looping over preallocated vectors.[2]

---

[2] You may wonder why dealing with column vectors is more efficient than dealing with row vectors. This has to do with the way that MATLAB stores arrays in memory. It is a useful general principle that you should prefer columns to rows when working in this language.

10.2.2 SPARSE MATRICES

In this book, we have already taken advantage of MATLAB's `sparse` commands. This group of commands allows us to work efficiently with matrices that contain a large number of zeros. `sparse` does this by instructing MATLAB to store only the non-zero elements of the matrix.

Sparsity can come in handy in many microeconometric calculations. Among others: when working with matrices containing observations on individual choice along a range of goods (which are mostly not chosen), transition matrices in dynamic choice (where individuals only transition between a relatively small number of the total states), and when working with identity matrices and their many off-diagonal zeros.

Fortunately, incorporating these matrices into your code is relatively simple, and can offer important speed-ups and memory savings. The following script demonstrates the benefits of these matrices in terms of memory space and time.

SparseTests.m

```
1  %----- (1) Set up test size, prefill results vector
2  N        = 1000;
3  Factor   = NaN(N,2);
4
5  %----- (2) Test memory improvement of sparse v
6  %             non-sparse -----
7  for i = 1:N
8      naive        = eye(i);
9      cool         = speye(i);
10
11     mem_n        = whos('naive');
12     mem_n        = mem_n.bytes;
13     mem_c        = whos('cool');
14     mem_c        = mem_c.bytes;
15
16     Factor(i,1)  = mem_n/mem_c;
17     clear naive cool
18 end
19
20 %----- (3) Test speed improvement of sparse v
21 %             non-sparse ------
22 for i = 1:N
23     naive    = eye(i);
24     cool     = speye(i);
25
26     tic; inv(naive); n=toc;
```

```
27    tic; inv(cool);   c=toc;
28
29    Factor(i,2) = n/c;
30    clear naive cool
31  end
32
33  %----- (4) Graphical output --------------------
34  subplot(2,1,1)
35  plot(1:N,Factor(:,1), 'LineWidth', 2)
36  xlabel('Size of Matrix (N\times N)',...
37        'FontSize', 10)
38  ylabel('Proportional Saving in Memory',...
39        'FontSize', 10)
40
41  subplot(2,1,2)
42  scatter(1:N, Factor(:,2), 'LineWidth',2)
43  xlabel('Size of Matrix (N\times N)',...
44        'FontSize', 10)
45  ylabel('Proportional Saving in Time',...
46        'FontSize', 10)
```

Figure 10.1 shows that the sparse commands can be very worthwhile. As the size of the initial (sparse) matrix grows, the storage size of sparse matrices is linearly more efficient compared to their non-sparse counterparts, while the relative efficiency in terms of operation time also increases approximately linearly.

Although sparse matrices must be created using the sparse command, MATLAB will operate on them as if they were a normal array, while respecting their sparse nature. For example, were we to call sum with a sparse matrix, this will return a sparse vector, with only as many entries as there are non-zero columns in the original matrix.

10.2.3 PROFILING CODE

In the function TimeTests, we introduced tic and toc to tell us how long the code takes run at various points. It is often useful to do this for an entire function to find the bottlenecks in our code. Fortunately, rather than having to introduce a series of counters at various points in our code, we can use the **profiler**. This essentially works as an ongoing series of tics and tocs. It allows us to see at each point of our code how long the code takes to run, and highlights for us those few places where bottlenecks exist, and where we may want to focus our attention.

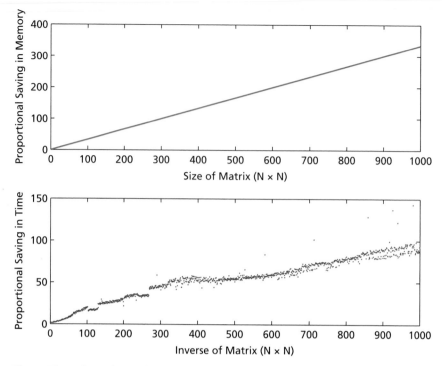

Figure 10.1 Why Bother with Sparsity?

We can run the profiler, for the sparse tests we ran in the previous section, and see what it tells us ...

```
>> profile on
>> SparseTests
>> profile off
>> profile viewer
```

The output of the profiler is shown in Figure 10.2.[3] It returns the run time at each step of the process, and highlights in red the few parts that hold up the script. Unsurprisingly, we see that most of the time is spent on inverting non-sparse matrices. If this were a real application, we could then consider optimizing this line (perhaps replacing it with sparse matrix operations).

10.2.4 WAITING ...

In the previous sections, we have worked under the assumption that our code can and should be further optimized. However, you will often find that even

---

[3] It also returns to us a line-by-line list of functions, but here we will just focus on the graphical output, and let you look through other output further if this is useful for you.

```
time   calls  line
           1    6 N     = 1000;
           1    7 Fxactor = NaN(N,2);
                8
                9 % ===================================================
               10 % (2) Test memory improvement of sparse matrix versus non-sparse as N increases
               11 % ===================================================
           1   12 for i = 1:N
0.27    1000   13   naive = eye(i);
<0.01   1000   14   cool = speye(i);
               15
<0.01   1000   16   mem_n = whos('naive');
        1000   17   mem_n = mem_n.bytes;
        1000   18   mem_c = whos('cool');
<0.01   1000   19   mem_c = mem_c.bytes;
               20
<0.01   1000   21   Factor(i,1)= mem_n/mem_c;
0.06    1000   22   clear naive cool
        1000   23 end
               24
               25 % ===================================================
               26 % (3) Test time improvement for sparse versus non-sparse
               27 % ===================================================
           1   28 for i = 1:N
1.06    1000   29   naive = eye(i);
0.02    1000   30   cool = speye(i);
               31
26.04   1000   32        tic; inv(naive); n=toc;
0.39    1000   33        tic; inv(cool); c=toc;
               34
<0.01   1000   35        Factor(i,2) = n/c;
0.38    1000   36        clear naive cool
        1000   37 end
```

Figure 10.2 MATLAB's Profiler Window

after writing your code as cleverly as possible, you are still forced to wait for a significant period of time while the code runs. If so, perhaps you would like to be updated in some way when MATLAB has completed its task. Fortunately, the inbuilt command `sendmail` can help. This allows you to send an email directly to your inbox from MATLAB, and even to include your MATLAB results as attachments.

The `sendmail` command is simple to use: it merely requires a recipient email address and, optionally, a subject, message, and attachments. However, there are a number of bugbears that must be ironed out if you wish to use password protected email. The function `MailResults` is a wrapper for `sendmail` that can be used to set your email preferences, and which you can use to notify yourself when your program finishes running.

MailResults.m

```
1  function MailResults(ResultsLabel, MatrixOut,...
2  YourEmail);
3  %-------------------------------------------------
4  % PURPOSE: Sends an email once containing
5  %          MATLAB's results
6  %-------------------------------------------------
7  % INPUTS: ResultsLabel : The name of the results
```

```
8   %                              matrix
9   %          MatrixOut    : A results matrix to send
10  %          YourEmail    : Your email address (in
11  %                         single quotes)
12  %---------------------------------------------------
13  % OUTPUT:  none
14  %---------------------------------------------------
15
16  %----- (1) Basic Email Preferences -------------
17  mail      = 'MyEmailAddress@MyMailProvider.com';
18  password  = 'MyPassword';
19
20  %----- (2) Advanced settings -------------------
21  setpref('Internet','SMTP_Server',...
22    'smtp.MyMailProvider.com');
23  setpref('Internet','E_mail',mail);
24  setpref('Internet','SMTP_Username',mail);
25  setpref('Internet','SMTP_Password',password);
26  props = java.lang.System.getProperties;
27  props.setProperty('mail.smtp.auth','true');
28  props.setProperty('mail.smtp.socketFactory.class',
29                  'javax.net.ssl.SSLSocketFactory');
30  props.setProperty('mail.smtp.socketFactory.port',
31                  '465');
32
33  %----- (3) Send email along with time spent
34  %              running program -------
35  MyToc    = toc;
36  dlmwrite('MatlabResults.csv', MatrixOut,...
37          'precision','%10.10f')
38  sendmail(YourEmail,ResultsLabel,...
39            strcat('toc = ', num2str(MyToc)),...
40            'MatlabResults.csv')
41
42  return
```

You can try this out by running the following at the MATLAB command line.

```
>> tic;
>> matrix = magic(10);
>> MailResults('magicmatrix', matrix,...
   'MyEmailAddress@MyMailProvider.com')
```

You will now have a magic square in your inbox!

10.3 **Parallel Computing**

In a nutshell, parallel computing allows for computationally intensive procedures to be separated and run in individual blocks rather than as one large job.
This is particularly useful in applications such as Monte Carlo simulation, or
other situations that rely on many processes that are independent from one
another.

Modern computers generally comprise a number of integrated processors.
These processors, called CPUs, can be thought of as the workhorse of the
computer when running calculations and simulations in MATLAB. You can
check how many cores your computer is using:

```
>> feature('numCores')

ans =
12
```

We can explicitly ask MATLAB to run our jobs using multiple CPUs—in many
situations this will substantially accelerate our code. What's more, programming this in MATLAB is remarkably easy, provided that you have the Parallel
Computing Toolbox installed in MATLAB. It is almost as simple as saying
parfor rather than for! The following code provides a brief example of this
and demonstrates some of the common pitfalls.

ParforBasics.m

```
1  %----- (1) Basic illustration of parfor ---------
2  matlabpool open
3  tic
4  parfor count = 1:12
5      count
6      pause(1)
7  end
8  toc
9
10 %----- (2) consider a problem that is not
11 %              parallelizable -----
12 fibonacci       = zeros(15, 1);
13 fibonacci(1)    = 1;
14 fibonacci(2)    = 1;
15
16 for c = 3:size(fibonacci, 1)
17     c
18     fibonacci(c)    = fibonacci(c - 1) +...
19                       fibonacci(c - 2);
20 end
```

```
21
22    %--- (3) this IS parallelizable, but needs
23    %            careful construction -----
24    MyObviousMatrix      = NaN(10, 2);
25    %MyObviousMatrix1     = NaN(10, 1);
26    %MyObviousMatrix2     = NaN(10, 1);
27
28    parfor count = 1:10
29        count
30        MyObviousMatrix(count, 1)    = 1;
31        MyObviousMatrix(count, 2)    = 2;
32
33    %MyObviousMatrix1(count)      = 1;
34    %MyObviousMatrix2(count)      = 2;
35    end
36    %MyObviousMatrix = [MyObviousMatrix1,
37    %MyObviousMatrix2];
38
39    return
```

The command `matlabpool open` tells MATLAB to operate across multiple cores. On our computer, we are using 12 cores. In block 1 of the code, we use a `parfor` loop, asking each loop to pause for one second. Were we to run this in a non-parallel fashion, it would, of course, take around twelve seconds. With a `parfor` loop, however, this took a mere one second: each `pause` is sent to a separate CPU.

In blocks (2) and (3), we illustrate some situations that *will not* work with parallel programming. `parfor` splits a loop between multiple cores so we cannot ask for information to be shared across iterations of the loop. In short, we require the problem to be 'parallelizable'—in the sense that it can be broken down into small blocks and then aggregated only once each block has finished. At each point in Fibonnaci sequence, we need to know the previous two realizations. For this reason the calculation is *sequential* rather than parallelizable.

There are trivially parallelizable problems in microeconometrics. Consider, for example, a bootstrap algorithm. This is perfect for a `parfor` loop. In each loop, we can calculate a subset of the bootstrap draws, and then once terminating all loops, pool these and calculate our final standard errors. Let's try this out with our familiar `auto.csv` data. `BootstrapOLS.m` is a script that estimates standard errors by bootstrapping a linear model. This illustrates the speed-up that can reasonably be achieved in parallel computations with some overhead.[4]

[4] If we wanted to compare this to Stata, we could type: `bootstrap, reps(10000): reg mpg price weight`

BootstrapOLS.m

```
1   clear
2   rng(1)
3
4   if matlabpool('size') == 0
5       matlabpool(12)
6   end
7
8   %-- (1) Open regression data --------------------
9   DataIn  = dlmread('auto.csv');
10  X       = DataIn(:, 2:3);
11  X       = [X, ones(74, 1)];
12  y       = DataIn(:, 1);
13  N        = size(X, 1);
14
15  [Beta, se]  = regress(y, X);
16  [Beta, se]
17
18  %-- (2) Bootstrap standard errors (100,000
19  %         draws) ----
20  reps             = 100000;
21  BootstrapBeta    = NaN(reps, 3);
22
23  tic
24  parfor count = 1:reps
25      MyIndex  = randi(N, N, 1);
26      BootX  = X(MyIndex, :);
27      BootY  = y(MyIndex, :);
28
29      BootstrapBeta(count, :) = [regress(BootY,...
30                                 BootX)]';
31  end
32  toc
33
34  [Beta, se, mean(BootstrapBeta)',...
35      std(BootstrapBeta)']
```

10.4 **Parallel Computing with the GPU**

We have just seen how MATLAB will let us take one big job, split it into a number of smaller jobs, and run these in parallel. This is a great way to save time—but you may want to push things further. In the previous section, we used 12 cores. Wouldn't it be cool if we could increase this to 20, 100, or even 1,000 cores?

It would not be feasible to put 1,000 CPUs in a personal computer. However, the Graphics Processing Unit comprises many mini-processors, each designed to render pixels in parallel. Luckily for people like us, these mini-processors can be tricked into working with numbers, rather than pixels, and can be used to parallelize microeconometric analysis. Parallel calculations on the GPU are now used in a large range of fields, including economics. See, for example, Aldrich et al. (2011) for an example of this type of computation.

MATLAB has built a large number of functions that take advantage of the GPU. To use these GPU functions, we require two things: (1) a GPU that works with these kinds of computations, and (2) MATLAB's Parallel Computing Toolbox. At the time of writing, this means that you must have access to an NVIDIA brand CUDA-enabled GPU. This is a very common type of GPU that you are likely to have in your computer—this can be checked at the MATLAB command line by typing gpuDevice. If your computer has both the toolbox and an NVIDIA GPU installed you should see something like the following excerpt:[5]

```
>> gpuDevice

ans =

    CUDADevice with properties:

                   Name: 'GeForce GT 630M'
                  Index: 1
      ComputeCapability: '2.1'
          SupportsDouble: 1
          DriverVersion: 5.5000
         ToolkitVersion: 5
```

If you have access to a computer with this hardware and software, the remainder is relatively easy. Much like MATLAB's sparse commands, its GPU commands are used by defining a different type of matrix, and then using normal MATLAB commands that recognize that we are dealing with GPU calculations. Essentially, we tell MATLAB to send a matrix to the GPU, run calculations on it there (in a parallel way), and then bring the processed information back from

---

[5] If you have the Parallel Computing Toolbox but do not have an NVIDIA GPU (or the appropriate software is not installed), you will see a message like: 'Error using gpuDevice (line 26). There is a problem with the CUDA driver associated with this GPU device.' If, however, you do not have the Parallel Computing Toolbox, you will see 'Undefined function or variable 'gpuDevice''.

the GPU to the CPU. The following code passes a vector of random draws to the GPU, takes the mean of each row, and brings it back:

```
>> N       = 10000;
>> GM      = gpuArray(rand(N));
>> tic;  Gmean = mean(GM);   G=toc
G =
      9.500e-04

>> Gback   = gather(Gmean);
```

Now, let's try the same thing without using the GPU. . .

```
>> CM      = rand(N);
>> tic;  Cmean = mean(CM);   C=toc
C =
      0.0439

>> ratio = C/G
ratio =
      46.4349
```

This very simple calculation suggests that we have managed to speed things up around 45 times by using the GPU (although this will vary a lot with the type of GPU used and the number of cores available). Finally, if we look at the types of these variables, we will see how our GPU arrays differ from normal variables:

```
>> whos
  Name         Size             Bytes   Class      Attributes

  C            1x1                  8   double
  CM           10000x10000  800000000   double
  G            1x1                  8   double
  GM           10000x10000        108   gpuArray
  Gback        1x10000          80000   double
  Gmean        1x10000            108   gpuArray
  N            1x1                  8   double
  Cmean        1x10000          80000   double
  ratio        1x1                  8   double
```

We now have a new class, gpuArray, for those matrices we passed to the GPU, whereas the values we brought back with the gather command are seen as normal vectors of double-precision numbers.

Before leaving GPU parallel programming, we feel it important to note that the suitability of the tool depends entirely upon the task assigned to it. The function GPUExample runs a relatively simple task: calculating $(X'X)^{-1}X'y$ on the GPU and on the CPU.

GPUExample.m

```
function Betas = GPUExample(N,k,method)
%------------------------------------------------
% PURPOSE: Creates and inverts matrices using GPU
%          computing
%------------------------------------------------
% INPUTS:   N         : The number of rows in the X
%                         and y matrices
%             k         : The number of columns in the
%                         X matrix
%             method : calculation method. Must be
%                         'GPU' or 'CPU'
%------------------------------------------------
% OUTPUT:  Betas   : result of inv(X'X)*(X'y)
%------------------------------------------------

if method=='GPU'
    X = gpuArray(rand(N,k));
    y = gpuArray(rand(N,1));
elseif method=='CPU'
    X = rand(N,k);
    y = rand(N,1);
end
tic
Betas = mldivide((X'*X),(X'*y));
toc
return
```

This function takes longer to run on the GPU than on the CPU.

```
>> GPUExample(1000,3,'GPU')
Elapsed time is 0.000845 seconds.

ans =

    0.3274
    0.2976
    0.3014
```

```
>> GPUExample(1000,3,'CPU')
Elapsed time is 0.000155 seconds.

ans =

    0.3274
    0.2976
    0.3014
```

10.5 **Other Tricks**

This chapter has, so far, been a tour through some 'low-cost' ways to speed up our MATLAB code. Before closing, we will point out a number of alternative tricks. However, due to their complex nature, we will leave you to explore these in your own time should you be interested.

First, you can use MATLAB's MEX-files. These files allow MATLAB to run code that has been written in Fortran or C/C++. These essentially act as 'plug-ins' to the MATLAB language. While the code is called fluidly from the MATLAB window, it retains the speed of the language in which it was originally written. However, as a word of caution, this requires learning the syntax of either Fortran or C/C++. While it is true that you will probably see performance enhancements in your code, this should be weighed against the cost of learning and implementing a new language.

Finally, you could consider online cloud computing. In this chapter, we have mentioned parallel coding using `parfor` and using your computer's GPU. However, both of these are inherently limited by the technology of your personal computer. Should you decide that you want to work on an extremely large number of cores, the recent advances in cloud computing offer a solution. Rather than running on the cores of your local machine, you can log in to a website that provides you with access to 'virtual cores' on which to run your code. The beauty of this infrastructure is that it places few limits on the number of cores which you can access—you simply use cores which live 'on the cloud', and then log off when you are finished. This is particularly useful if you will be doing large one-off jobs that require high computational resources for a short period of time. Two options are the Amazon EC2,[6] and PiCloud. Both of these are available for use with MATLAB and, at least for the case of Amazon's EC2, come with documentation on the MATLAB website.

---

[6] Amazon EC2, or Amazon Elastic Compute Cloud is a leader in cloud computing and provides computational resources which can be rented by the hour.

11 ... and Slowing Things Down

> Now this is not the end. It is not even the beginning of the end.
> But it is, perhaps, the end of the beginning.
>
> Winston Churchill*

Time to say goodbye. In this book, we have introduced the basics of MATLAB—not as an end in itself but as a means to discuss interesting ideas, models, and methods in theory-based empirical analysis. Time, then, for you to apply these ideas to your own research. Here—in general terms—is how we suggest doing so.

(i) *Find a real research question.* Hypotheticals are great for learning—as is simulated data—which is why we have used them so extensively in this text. Now it is time to apply what you have learned to a real research question and a real dataset. This, ultimately, is where lies the real challenge of theory-based empirical analysis: in modelling the messiness of the real world, where the lurks and perks of specific empirical contexts often do not align perfectly with textbook models.

(ii) *Write a model, and code it using MATLAB functions.* Keep it simple; you can always complicate matters later.

(iii) *Decide how you want to estimate your model* (Maximum Likelihood, Maximum Simulated Likelihood, GMM, Linear Programming, etc.). Write an objective function, and code it using MATLAB functions.

(iv) *Simulate, simulate, simulate.* Simulate to check your code; simulate to understand the behaviour of your model; simulate large datasets and estimate on that simulated data, to confirm that you can recover your model parameters.

(v) *Estimate—on real data, at last.* This should feel like a cautious, halting approach—at least initially. You should avoid grandiose temptations—the temptation to run your estimator once and then start reporting results. Rather, you might want to check that your estimator returns sensible results in some constrained region of the parameter space, then gradually allow for a wider search. You may want to return to your simulated code—to confirm that your model can be simulated and estimated with the parameters that your estimator is returning from the real data.

(vi) Rethink. Refine. Revise. *Repeat.*

---

* Churchill, W., *The End of the Beginning*, The Lord Mayor's Luncheon, Mansion House (1942).

If you're not having fun, you're doing something wrong.

Marx

BIBLIOGRAPHY

Acemoglu, D. (2008). *Introduction to Modern Economic Growth*. Princeton, New Jersey: Princeton University Press.

Adda, J., and Cooper, R. (2000). Balladurette and Juppette: A Discrete Analysis of Scrapping Subsidies. *Journal of Political Economy, 108*(4), 778–806.

Adda, J., and Cooper, R. (2003). *Dynamic Economics*. Cambridge, Massachusetts: MIT Press.

Aldrich, E. M., Fernández-Villaverde, J., Ronald Gallant, A., and Rubio-Ramírez, J. F. (2011). Tapping the Supercomputer Under Your Desk: Solving Dynamic Equilibrium Models with Graphics Processors. *Journal of Economic Dynamics and Control, 35*(3), 386–93.

Bajari, P., Hong, H., Krainer, J., and Nekipelov, D. (2010). Estimating Static Models of Strategic Interactions. *Journal of Business & Economic Statistics, 28*(4).

Bellman, R. (1957). *Dynamic Programming*. Princeton, New Jersey: Princeton University Press.

Berry, S. T. (1992). Estimation of a Model of Entry in the Airline Industry. *Econometrica*, 889–917.

Bond, S., and Söderbom, M. (2005). Adjustment Costs and the Identification of Cobb Douglas Production Functions. Economics Papers 2005-W04, Economics Group, Nuffield College, University of Oxford.

Bowman, A. W., and Azzalini, A. (1997). *Applied Smoothing Techniques for Data Analysis: The Kernel Approach with S-Plus Illustrations*. Oxford: Oxford University Press.

Bresnahan, T. F., and Reiss, P. C. (1990). Entry in Monopoly Markets. *The Review of Economic Studies, 57*(4), 531–53.

Bresnahan, T. F., and Reiss, P. C. (1991a). Empirical Models of Discrete Games. *Journal of Econometrics, 48*(1), 57–81.

Bresnahan, T. F., and Reiss, P. C. (1991b). Entry and Competition in Concentrated Markets. *Journal of Political Economy*, 977–1009.

Cameron, A. C., and Trivedi, P. K. (2005). *Microeconometrics: Methods and Applications*. New York: Cambridge University Press.

Casella, G., and Berger, R. L. (2002). *Statistical Inference*. Pacific Grove, California: Duxbury Press/Thomson Learning, 2 edn.

Ciliberto, F., and Tamer, E. (2009). Market Structure and Multiple Equilibria in Airline Markets. *Econometrica, 77*(6), 1791–828.

Collard, F. (2013). Dynamic Programming. URL http://fabcol.free.fr/pdf/lectnotes7.pdf.

de Paula, A., and Tang, X. (2012). Inference of Signs of Interaction Effects in Simultaneous Games with Incomplete Information. *Econometrica, 80*(1), 143–72.

Delgado, M. A., and Robinson, P. (1992). Nonparametric and Semiparametric Methods for Economic Research. *Journal of Economic Surveys, 3*, 201–49.

Dixit, A. K. (1990). *Optimisation in Economic Theory*. Oxford: Oxford University Press.

Eisenhauer, P., Heckman, J. J., and Mosso, S. (2014). Estimation of Dynamic Discrete Choice Models by Maximum Likelihood and the Simulated Method of Moments. NBER Working Paper 20622, National Bureau of Economic Research.

Fafchamps, M., McKenzie, D., Quinn, S., and Woodruff, C. (2014). Microenterprise Growth and the Flypaper Effect: Evidence from a Randomized Experiment in Ghana. *Journal of Development Economics, 106*, 211–26.

Fan, J., and Gijbels, I. (1996). *Local Polynomial Modelling and its Applications*. London: Chapman and Hall.

Friedman, J. (1984). A Variable Span Scatterplot Smoother. Tech. Rep. 5, Stanford University.

Geweke, J., Keane, M. P., and Runkle, D. (1994). Alternative Computational Approaches to Inference in the Multinomial Probit Model. *The Review of Economics and Statistics*, 76(4), 609–32.

Gole, T., and Quinn, S. (2014). Committees and Status Quo Bias: Structural Evidence from a Randomized Field Experiment. University of Oxford Department of Economics Discussion Paper, No. 733.

Gourieroux, C., and Monfort, A. (1997). *Simulation-Based Econometric Methods*. Oxford: Oxford University Press.

Grieco, P. L. (2014). Discrete Games with Flexible Information Structures: An Application to Local Grocery Markets. *The RAND Journal of Economics*, 45(2), 303–40.

Hahn, B., and Valentine, D. (2013). *Essential MATLAB for Engineers and Scientists*. Waltham, Maryland: Academic Press/Elsevier, 5 edn.

Hall, A. R. (2005). *Generalized Method of Moments. Advanced Texts in Econometrics*. Oxford: Oxford University Press.

Hansen, L. P. (1982). Large Sample Properties of Generalized Method of Moments Estimators. *Econometrica*, 50(4), 1029–54.

Härdle, W. (1991). *Applied Nonparametric Regression*. Cambridge: Cambridge University Press.

Härdle, W., and Linton, O. (1994). *Handbook of Econometrics*, vol. 4, chap. Applied Nonparametric Methods, pp. 2295–339. Amsterdam: Elsevier.

Härdle, W., Hall, P., and Ichimura, H. (1993). Optimal Smoothing in Single-Index Models. *Annals of Statistics*, 21(1), 157–78.

Härdle, W., Werwatz, A., Müller, M., and Sperlich, S. (2004). *Nonparametric and Semiparametric Models*. Berlin Heidelberg: Springer.

Hastie, T., and Tibshirani, R. (1990). *Generalized Additive Models*. London: Chapman and Hall.

Heckman, J. J. (1978). Dummy Endogenous Variables in a Simultaneous Equation System. *Econometrica*, 931–59.

Heckman, J. J., Ichimura, H., and Todd, P. (1998). Matching as an Econometric Evaluation Estimator. *Review of Economic Studies*, 65, 261–94.

Hodges, J. L., and Lehmann, E. L. (1956). The Efficiency of Some Nonparametric Competitors of the t-Test. *Annals of Mathematical Statistics*, 27, 324–35.

Horowitz, J. (1992). A Smoothed Maximum Score Estimator for the Binary Response Model. *Econometrica*, 60, 505–31.

Ichimura, H. (1993). Semiparametric Least Squares (SLS) and Weighted SLS Estimation of Single-Index Models. *Journal of Econometrics* 58, 71–120.

Ichimura, H., and Todd, P. (2007). *Handbook of Econometrics*, vol. 6B, chap. Implementing Nonparametric and Semiparametric Estimators, pp. 5369–468. Elsevier.

Judd, K. L. (1998). *Numerical Methods in Economics*. Cambridge, Massachusetts: MIT Press.

Keane, M., and Wolpin, K. (1994). The Solution and Estimation of Discrete Choice Dynamic Programming Models by Simulation and Interpolation: Monte Carlo Evidence. *The Review of Economics and Statistics*, 76(4), 648–72.

Keane, M. P., and Wolpin, K. I. (1997). The Career Decisions of Young Men. *Journal of Political Economy*, 105(3), 473–522.

Klein, R., and Spady, R. (1993). An Efficient Semiparametric Estimator for Binary Response Models. *Econometrica, 61*, 387–421.

Laibson, D. (1997). Golden Eggs and Hyperbolic Discounting. *Quarterly Journal of Economics, 112*(2), 443–77.

Li, Q., and Racine, J. (2006). *Nonparametric Econometrics: Theory and Practise*. Princeton, New Jersey: Princeton University Press.

Linton, O., Chen, R., Wang, N., and Härdle, W. (1997). An Analysis of Transforms for Additive Nonparametric Regression. *Journal of the American Statistical Association, 92*, 1512–21.

Ljungqvist, L., and Sargent, T. J. (2000). *Recursive Macroeconomic Theory*. Cambridge, Massachusetts: MIT Press, 2 edn.

Loftsgaarden, D. O., and Queenberry, C. P. (1965). A Nonparametric Density Function. *Annals of Mathematical Statistics, 36*, 1049–51.

Manski, C. (1975). Maximum Score Estimation of the Stochastic Utility Model of Choice. *Journal of Econometrics, 3*, 205–28.

McFadden, D. (1974). The Measurement of Urban Travel Demand. *Journal of Public Economics, 3*(4), 303–28.

McFadden, D. L. (2000). Economic Choices. Nobel Prize in Economics documents 2000–6, Nobel Prize Committee.

Michie, D. (1968). 'Memo' Functions and Machine Learning. *Nature, 218*(5136), 19–22.

Morris, S., and Shin, H. S. (2003). Global Games: Theory and Applications. *Econometric Society Monographs, 35*, 56–114.

Nadaraya, E. A. (1964). On Estimating Regression. *Theory of Probability and Its Applications, 9*, 141–2.

Pagan, A., and Ullah, A. (1999). *Nonparametric Econometrics*. Cambridge: Cambridge University Press.

Parzen, E. (1962). On the Estimation of a Probability Density Function and Mode. *Annals of Mathematical Statistics, 33*, 1065–76.

Powell, J. (1994). *Handbook of Econometrics*, chap. Estimation of Semiparametric Models, pp. 291–316, 75. Amsterdam: Elsevier.

Powell, J., Stock, J., and Stoker, T. (1989). Semiparametric Estimation of Index Coefficients. *Econometrica, 57*(6), 1403–30.

Ramsey, F. P. (1928). A Mathematical Theory of Saving. *The Economic Journal, 38*(152), 543–59.

Robinson, P. M. (1988). Root-N-Consistent Semiparametric Regression. *Econometrica, 56*(4), 931–54.

Rosenblatt, M. (1956). Remarks on Some Nonparametric Estimators of a Density Function. *Annals of Mathematical Statistics, 27*, 832–7.

Rust, J. (1994a). Estimation of Dynamic Structural Models, Problems and Prospects: Discrete Decision Processes. Proceedings of the 6th World Congress of the Econometric Society. Cambridge University Press.

Rust, J. (1994b). *Handbook of Econometrics*, vol 4, chap. Structural Estimation of Markov Decision Processes, pp. 3081–143. Amsterdam: Elsevier.

Rust, J. (2000). Parametric Policy Iteration: An Efficient Algorithm for Solving Multidimensional DP problems. Mimeo, University of Maryland.

Sargent, T., and Stachurski, J. (2013). Quantitative Economics. URL http://quant-econ.net/.

Silverman, B. (1986). *Density Estimation for Statistics and Data Analysis*. London: Chapman and Hall.

Stachurski, J. (2009). *Economic Dynamics: Theory and Computation*. Cambridge, Massachusetts: MIT Press.

Stoker, T. (1986). Consistent Estimation of Scaled Coefficients. *Econometrica*, *54*, 1461–81.

Stokey, N., and Lucas, R. (1989). *Recursive Methods in Economic Dynamics*. Cambridge, Massachusetts: Harvard University Press.

Sweeting, A. (2009). The Strategic Timing Incentives of Commercial Radio Stations: An Empirical Analysis Using Multiple Equilibria. *RAND Journal of Economics*, *40*(4), 710–42.

Tamer, E. (2003). Incomplete Simultaneous Discrete Response Model with Multiple Equilibria. *Review of Economic Studies*, *70*, 147–65.

Todd, P., and Wolpin, K. I. (2006). Assessing the Impact of a School Subsidy Program in Mexico: Using a Social Experiment to Validate a Dynamic Behavioral Model of Child Schooling and Fertility. *American Economic Journal*, *96*(5), 1384–417.

Train, K. E. (2009). *Discrete Choice Methods with Simulation*. Cambridge: Cambridge University Press.

Wan, Y., and Xu, H. (2014). Semiparametric Identification of Binary Decision Games of Incomplete Information with Correlated Private Signals. *Journal of Econometrics*, *182*(2), 235–46.

Watson, G. (1964). Smooth Regression Analysis. *Sankhya Series A*, *26*, 359–72.

Wolpin, K. I. (1984). An Estimable Dynamic Stochastic Model of Fertility and Child Mortality. *Journal of Political Economy*, *92*(5), 852–74.

Wolpin, K. I. (2013). *The Limits of Inference Without Theory*. Cambridge, Massachusetts: MIT Press, 2 edn.

Wooldridge, J. M. (2010). *Econometric Analysis of Cross Section and Panel Data*. MIT Press, 2 edn.

Xu, H. (2014). Estimation of Discrete Games with Correlated Types. *The Econometrics Journal*, *17*(3), 241–70.

Yatchew, A. (2003). *Semiparametric Regression for the Applied Econometrician*. Cambridge: Cambridge University Press.

■ INDEX